Structural Design in Building Conservation

No building is properly conserved if it is not structurally sound. Consequently architects, engineers and conservation officers need an adequate grounding in the technology as well as in the materials and the historic origins of the building.

Structural interventions to historic buildings are however an integral part of the effort to select and update their design and their historic and cultural values. *Structural Design in Building Conservation* deals with such design issues and shows how technical choices integrate with the planning and architectural outcomes in a conservation project. It brings together theory with current conservation technology, discussing the possibilities of structural details and strategies in architectural expression, and is particularly directed at students of architectural conservation technology and practising engineers and architects.

Central to this outline is the discussion of case studies, which is organised around such themes as addition of roofs, requalification of space, strengthening and re-use of fabric, repristination, additions, completions, stiffness adjustments and the correction of past mistakes etc, and the reader is encouraged to appraise directly the solutions.

The book introduces the philosophy of structural interventions within the range of conservation theories and discusses practices in various European countries. It then outlines the main types of strengthening, repairs and interventions in terms of building types and the structural nature of the main elements to be strengthened (linear structures, frames, plates and shells). Significant case studies are presented that cover a very wide range of historic types and conversions, not only monumental masonry structures like neoclassical buildings, major temples, churches, public buildings and museums, but also more utilitarian structures like historic mills, early reinforced concrete structures and vaulting types.

Dimitris Theodossopoulos is Lecturer in Architectural Technology and Conservation at the University of Edinburgh, UK. He trained as a civil engineer and is a specialist in architectural conservation.

Structural Design in Building Conservation

Dimitris Theodossopoulos

Routledge
Taylor & Francis Group

LONDON AND NEW YORK

First published 2012
by Routledge
2 Park Square, Milton Park, Abingdon, Oxon OX14 4RN

Simultaneously published in the USA and Canada
by Routledge
711 Third Avenue, New York, NY 10017

Routledge is an imprint of the Taylor & Francis Group, an informa business

British Library Cataloguing in Publication Data
A catalogue record for this book is available from the British Library
Library of Congress Cataloging in Publication Data
Theodossopoulos, Dimitris.
Structural design in building conservation/Dimitris Theodossopoulos.
    p. cm.
  Includes bibliographical references and index.
  1. Historic buildings—Conservation and restoration.
2. Architecture–Conservation and restoration. I. Title.
  TH3301.T54 2012
  690'.24—dc23   2011040939

ISBN: 978–0–415–47945–5 (hbk)
ISBN: 978–0–415–47946–2 (pbk)
ISBN: 978–0–203–88725–7 (ebk)

Typeset in Univers
by Swales & Willis Ltd, Exeter, Devon

Contents

Contents

Preface

This book aims to introduce the reader to the structural issues and the process of making choices involved in major design interventions to historic architecture, so these are addressed to architects, engineers or conservation practitioners at the early stages of their careers or later stages of their education. The discussion moves beyond the conservation of the building fabric or archaeological evidence towards those projects that seek to present or update the historic and cultural values of these buildings, even by providing new uses or new space configurations. This often appears as removal and alteration of existing fabric or addition of new elements, but in truth the discussion is about those critical acts that permit the understanding of building fabric with historic, artistic or cultural values and the transmission of these values to this and future generations.

Education is fundamental in this sector as only a limited amount can be learned through practice and the learning experience of taking a course permits the right focus and eventually creates the sensitivity such buildings require. The need for such a book became evident to me through my teaching at undergraduate and postgraduate levels in architecture and structural engineering, especially in design projects and dissertations. Because of the syllabus, it is inevitable that the essential steps in dealing with existing fabric (historic construction technology and its culture, conservation theory, building survey, historiography, specialist legislation and management practice) are covered only briefly. But, often, even in specialist courses there is limited time for these essential steps to link with the more complex process of decision-making and explorations that design in historic buildings requires. The range of projects presented is expected to demonstrate that conservation goes far beyond a linear application of techniques or the repristination of a specific past configuration.

In this process of integration to increasing complexity, it is necessary to review initially the range of interventions that usually affect historic buildings, including even those that have disregarded their historic, artistic and architectural values. This discussion is supported by an outline of conservation theories that focus on the treatment of the fabric as a source of historic or artistic evidence. No attempt has been made to offer definitions or discuss the differences between them (preservation, conservation, retrofitting, etc.), but some terms that I find pointless or offensive to a historic building, such as adaptive reuse, are omitted altogether. Rather than providing stiff definitions or a prescriptive approach, theories, regulations and charters are discussed as the cultural luggage of the operator who then has to make the critical decisions on the integration of the new intervention with the character of the building.

Current theory of structures and mechanics cannot, of course, apply to historic fabric in the same precise manner as in the design of modern buildings, but designers often work in the preliminary design of a structure using conceptual schemes of behaviour. Classifications to building types were deemed necessary here as well (cellular vs. frame structures, vaults, etc.), as also was the study of their behaviour with regards to standardised actions (dead load, wind, earthquake) where the performance of the new interventions needs to be certified. Assessment of the behaviour of the existing building has also specific needs within this context which need to be highlighted as part of the preparation process.

It is often the case that incorporating a new structure requires strengthening or upgrading of the load-bearing capacity of the existing structure or specific ways of linking their performance or completely detaching them. A range of techniques is briefly presented that are relevant to the problems.

The contribution of this book is in highlighting how these essential steps build towards a critical dialogue and updating of historic architecture to values that can be shared among generations. A range of design interventions has been selected in Chapter 6 according to the most frequently encountered structural problems, which often do not correspond directly to the architectural priorities. On the contrary, the case studies may have a strong component of structural resolution that has affected the architectural choices, either through preservation or enhancement of the building or site. In this key part of the book, the applications of the more theoretical earlier chapters become evident as does the process where the needs of the existing fabric are innovatively treated or combined with those of the proposed one.

Each area is introduced with a summary of the technical issues and the way they often affect the design procedure. A major case study is then examined in more detail following an outline that encourages the readers to understand the options available to the designers and validate the solutions: the historic and material conditions of the existing fabric are presented in terms of the design intentions and the possible strategies are discussed, guiding the reader to evaluate the actual solution through alternatives or critical enquiry.

I believe this is a fundamental approach towards the design of the most respectful yet architecturally successful intervention to a historic building. The designers, architect or engineer, work in liaison with the client, contractors and regulatory authorities and have a professional role to balance their needs and abilities in order to deliver their design. They also represent the aspirations and attitudes of the community and they have not only to capture the values the building or monument represents for them, but also to express them through the intervention in a clear and sustainable manner and transmit them to future generations. The designer, therefore, needs a genuine confidence that only the knowledge of the monument and its conditions can provide, backed by a will to actively look for adequate solutions. It is the aim of this book to promote the development of such skills by highlighting the challenges and the successful, or not so successful, battles of designers and their relationships with the essence of the material in their hands.

I hope the reader will place this book within other efforts to inspire designers to develop their own critical and dialectic approach to historic buildings, with respect to their design characteristics and history of human interactions with them. Past books on the subject, such as those by S. Cantacuzino (1975), C. Schittich (2003), Cramer and Breitling (2007), or the carefully illustrated and compiled *Atlante del Restauro* by my professor, G. Carbonara (2005), have opened the way towards a synthesis of all the disciplines around conservation and the reader is strongly encouraged to read this book as a continuity of those efforts.

Similarly, there are textbooks that deal with the technical issues in much more depth than the space of this book permits (Beckmann and Bowles 2004; Croci 1998, 2001; Robson 1999; Forsyth 2007a, 2007b; Alvarez and Gonzalez 1994; CIRIA 1994; Ashurst 1998; etc.), and the design potential of the techniques described there is explored in the chapters that offer introduction to the case studies. These are summarised in Chapter 3 and the presentation highlights what I believe, from my own experience, an engineer or architect who starts in conservation needs to know. At the same time, the discussion of structural issues revolves around contemporary debates which recognise the stronger inter-relationship with environmental performance, the increased sensitivity of regulations to disproportionate collapse and fire, the links with materiality, a balanced functionalism and, essentially, a clear expression of the role of structure in the character of a building.

Access to Italian, Spanish and Greek sources, I hope, is a great benefit for the reader and not a limitation. It has often been a problem in my teaching to refer to examples with an advanced theoretical or technical approach that are not easily available in the English language. I hope to use in the future more examples from Germany, the Americas or China that can enrich the cultural context of designing with historic fabric (Stubbs *et al.* 2011). It is inevitable, however, that I should draw many examples from Italy, and specifically Rome, as through my studies there I became familiar with the rich stratification and cultural interactions of the monuments and sites there. It is essential for anyone who starts working in this field to engage with such strong examples as they highlight complexity but also the wide range of choices in their conservation, by operators from these and past generations. It is also inevitable that many other examples come from

Greece, my country of origin, and Spain, where my wife Cristina comes from (especially the North, Asturias and its capital Oviedo). Many other examples also come from Nottingham and Edinburgh where I have developed my academic career.

Thanks

Appreciation of these issues throughout my work could not have been possible without the essential education from my professors in the field, and in particular those in the Sapienza in Rome, Professor Giorgio Croci, Professor Giovanni Carbonara and Dr Alberto Viskovic, as well as Prof. Dionysios Verras in Patras. They have all been inspirational figures as educators, critics and professionals in the field. My supervisors in my PhD in Edinburgh (Prof. Braj Sinha, Prof. Asif Usmani and Prof. Angus Macdonald) are also warmly acknowledged for helping me set my enthusiasm for historic structures in a truly scientific context.

Many colleagues and friends are thanked for their direct and indirect contribution to the discussions in this book as well as for valuable material from their own work. In no particular order these are Konstantinos Karanassos for showing us his precise work at the Propylaia of the Athenian Acropolis, Donato Abruzzese for his passion in conservation education among engineers, Renzo Chiovelli for his calm demonstrations of the value of conservation theory, Beatrice Vivio for her insight into Franco Minissi's work, my colleague Miles Glendinning for our collaboration at his MSc in Architectural Conservation in Edinburgh.

A special mention should go to John Barber and the team at AOC Archaeology in Edinburgh for introducing me to the Scottish brochs and making me 'lose myself to the kindness of the anonymity of prehistory'. I hope this work gives a glimpse to the character and conservation of these unique structures.

My work in the field as well as the preparation of the book put me in contact or collaboration with some unique operators in Edinburgh, including Ted Ruddock, Bob Heath, John Addison and Douglas Johnston, and in England, such as Poul Beckmann and James Sutherland. I have to acknowledge my inspiration from Prof. Charles McKean at Dundee University and, in particular, a lecture where he discussed the intervention at Koldinghus Castle, maybe this whole book started indirectly from that point. Special mention should be made to my friends in Nottingham Civic Society, Hilary Silvester and Ian Wells, for their unique insight into conservation problems in the area and for introducing my wife and me to some incredible sites.

The preparation of the book was also an opportunity to get to know better some projects and the work of some exceptional European architects and engineers such as Angela Garcia de Paredes in Madrid, Francisco Partearroyo and his team in Arquimatica also in Madrid, Inger and Johannes Exner in Denmark, form-TL in Germany, GMP von Gerkan Marg und Partner in Hamburg, Estudio Primitivo Gonzalez in Valladolid, Geoff Clifton and

Gifford in England, Buro Happold and, in particular, Iain Hargreaves in their Edinburgh office, Prof. Andrea Bruno in Turin.

Knowledge of a wide range of case studies became possible by the enthusiastic help of Margarita Escotet, the librarian of the Colegio Oficial de Arquitectos de Asturias COAA in Oviedo, as well as the valuable discussions of both design and professional aspects, and generous permissions by the technical magazines *Architects' Journal, New Civil Engineer and Structural Engineer*. The libraries at the School of Architecture (ETSAM) and the Colegio de Ingenieros CICCP, both in Madrid, are also warmly thanked for their support and the very useful recording of architecture and engineering history in their collections.

Access to some key monuments became possible with the generous support of many state agencies, particularly the Soprintendenza Archeologica di Roma and TANA (Hellenic Archaeological Receipts Fund).

A significant number of the issues presented here have been either tested in my lectures or explored through teaching models and some of my students in Edinburgh who made particular contributions need to be thanked, including Adam Neep, Laura Barr, Lee Kynoch, Emma Garland, Shona Black, Melinda Jin, Weifeng Kong, Lynne Mackay, Marietta Galazka, Alexandra Kuklinski, Jamie Henry and Xuhong Zheng. Our work would not have been possible without the innovative and enthusiastic support of our technicians Alistair Craig, Malcolm Cruickshank and Rachel Travers. Special thanks to my friend Andy Jones for his invaluable help with my original structural models for the vaults in Holyrood Abbey.

Many other professionals and academics have to be thanked for their advice: R. Barthel, O. Rio Suarez, R. Ousterhout, J. Crow, D. Athanasoulis, S. Mamaloukos, M. Sakellariou, C. A. Syrmakezis, D. Beskos, P. Lourenço, P. Roca, S. Huerta, J. Ochsendorf. Inevitably some might have been missed.

I have also to thank sincerely Taylor & Francis for the opportunity to write this book, and particularly all those people who believed in this project, gave me encouraging and critical comments and showed immense patience, including Brian Guerin, Alice Aldous, Sarah Olney, Tony Moore and Simon Bates. I also need to thank those people who supported the production of this book, in particular Tamsin Ballard, Rachel Hutchings and the indexer Sue Lightfoot.

I need also to acknowledge the support of the School of Architecture at the University of Edinburgh for the sabbatical teaching leave which allowed the bulk of the book to be investigated and produced.

Finally, I offer the most important thanks to my wife Cristina González-Longo, an architect and teacher also specialising in conservation. Without her love, patience, sacrifice, insight from professional practice, critical eye, personal encouragement and faith in my project this book would not have been possible. I also hope the book can transmit some of the balanced and measured approach to things that my parents gave me through their love and education.

Acknowledgements

In addition to the persons thanked in the Preface, I would like to acknowledge various other people, companies or agencies that have very kindly allowed me to use material from their work for this book and even gave me their support and comments in the preparation of the case studies.

ΤΑΠΑ (Hellenic Archaeological Receipts Fund) for Figures 1.3 and 6.29; Mrs Margarita Escotet, librarian of Colegio Oficial de Arquitectos de Asturias (COAA) and the Library of COAM for access to publications of valuable Spanish case studies and Figure 1.4; *New Civil Engineer* magazine (especially James McCarthy) for the up-to-date information on current projects in Figures 3.24 and 6.65 and in my diagrams (1.7); Teatro Regio di Parma and Foto Studio Carra for Figure 1.8; the library of the School of Architecture (ETSAM) of the Technical University of Madrid (UPM), especially the director Mrs Blanca Ruilope Urioste, for the use of images from their rare books collection in Figures 2.1, 3.2, 3.9, 3.14, 3.16, 3.26 and 3.27; special thanks to Prof. Santiago Huerta at the ETSAM for creating the online library of sources in the History of Construction that allowed me access to such valuable historic information; the Scottish Lime Centre for allowing me and my students access to their Penicuik House (Figure 2.3); Prof. Paulo Lourenço for allowing me to use his diagrams in Figures 2.5 and 5.10; the Italian Ministero dei Beni e Attivitá Culturali and Soprintendenza Speciale per i Beni Archeologici di Roma for kind permission to use photos from the very important monuments under their care in Figures 3.1, 6.24, 6.25 and 6.56; Prof. Braj Sinha for the diagram in Figure 3.6; Dr Isabelle Titeux and her co-authors for Figure 3.7; Mr James Sutherland for both his diagram in Figure 3.13 and his encouragement; Arup and the editor of the *Arup Journal*, Mr David J. Brown, for Figures 3.18 and 6.60; Mr Bob Heath and Edinburgh World Heritage Trust for the seminal drawing in Figure 3.20; *Architects' Journal* (especially Mr Simon Hogg) for allowing me to reproduce analytical diagrams from key contemporary buildings in Figures 3.23, 6.3, 6.5, 6.10 and 6.55; Madeleine Donachie from the *American Journal of Archaeology* for Figure 3.32 and valuable copyright advice; D. Juan Antonio Becerril Bustamante, director of the Revista de Obras Públicas, Colegio de Ingenieros de Caminos, Canales y Puertos for Figure 3.33; Izaskun Fernández de Castillo from Fundación Catedral Santa María and Quintas Fotógrafos for the photos from Vitoria Cathedral in

Acknowledgements

Figures 3.34 and 6.45; Rev. Richard Spray for access to Blyth Priory that enabled the survey in Figure 4.1; John Barber, Andy Heald, Graeme Cavers and Gemma Hudson from AOC Archaeology for permitting use of Figures 4.2, 4.3 and 4.6; CONTROLS srl for an image of their services in Figure 4.4; Lincoln Cathedral for the data in Table 4.1; Dr Alberto Viskovic for Figure 5.15 and, together with Studio Croci, for Figures 5.3 and 6.32; CIRIA for the reproduction of images from their publications in Figures 5.4, 5.8 and 5.12; Sika Corp for illustration of their products in Figure 5.6.

The breadth of the case studies and the discussion of their high qualities in Chapter 6 would not have been possible without the generous contribution of many of their authors, especially for those that I got to know mainly through literature resources. The following architects, engineers, clients or agencies are thanked: gmp (von Gerkan, Marg and Partners Architects) Bettina Ahrens and Hans Georg Esch for Figures 6.1 and 6.2; the National Maritime Museum in Greenwich for permission to use Figure 6.4; Buro Happold and Robert Greshoff for Figure 6.6; Arquimática SPL (Francisco Partearroyo) for their drawings in Figures 6.7 and 6.9; the City of Madrid for permission for Figure 6.8; the Deutscher Bundestag for permission for Figure 6.13; Dr Beatrice Vivio and Archivio Minissi for Figures 6.14 and 6.15; dott.sa Claudia Del Monti for her diagram in Figure 6.16; Diputación de Palencia for Figure 6.17; Prof. Angela Garcia de Paredes for details from the Paredes Pedrosa Arquitectos roof over the Roman Villa of La Olmeda in Figure 6.18; the engineering firm form-TL for the image of their project in Malta in Figure 6.20; Dr Miles Glendinning for his photos in Figures 6.22 and 6.36; Dr K. Karanassos and YSMA (Acropolis Restoration Service) for his drawing in Figure 6.30; Musei Civici di Padova for permission for Figure 6.34; Deutsche Fotothek for Figure 6.35; the Concrete Centre and Graham Bizley for the diagram in Figure 6.38; Inger and Johannes Exner for allowing me to reproduce material from their work in Koldinghus in Figures 6.39 and 6.40; Prof. A. Bruno for the drawing from Palazzo Carignano in Figure 6.42; Western Pennsylvania Conservancy for Figure 6.43; Cristina González-Longo for her drawing in Figure 6.51; Geoff Clinton and Gifford for their drawing in Figure 6.52 and valuable comments; Estudio Primitivo Gonzalez Arquitectos for Figures 6.61 and 6.62, and their suggestions.

Chapter 1

Theoretical aspects of structural interventions

Many of the approaches that are represented by the areas of intervention discussed in this book clearly stem from a specific necessity (create a new enclosed space, strengthen a ruin, remove internal supports). Other interventions have to do with the preservation and presentation of the fabric and fall into clearer definitions of conservation. The key question in such projects is always 'what to conserve' and the answer does not always have to do with the building matter itself.

Whatever its scale or scope, an intervention starts eventually as a critical act of a similar nature to design (Bonelli 1963), which weighs the values the historic building will carry forward, resulting in their more fruitful integration in the project and a richer interrogation. Reflection on current conservation theories, therefore, is necessary at the decision-making process and widens the range of choices. Study of their evolution and application to structural design in this chapter will search the synthesis of the purely architectural aspects with the strictly materialistic ones, attempting a sketch of a theory pertinent to building fabric and structural issues. Compared to architectural design, structural design and technical advancements have followed different rhythms and processes throughout history (Ousterhout 2008), which may result in different, or even conflicting, values to be conserved. Similarly, the structural project might have very different priorities, e.g. an addition or removal could cause disproportionate complex interactions with other parts of the building.

However, it should become clear that emphasis on the structural aspects has to do mainly with professional divisions of the discipline or specific design or construction skills rather than a different conceptual approach. Structure occasionally contains prevailing values of its own (Charleson 2005) but this should not divert from the fact that it is an integral part of the architectural organism and, even at its most utilitarian aspect (public infrastructure like viaducts or dams), it is an essential part of our built environment.

This review will also show that, at least in Europe, philosophy and practice also differ between countries. This makes even the regulatory, professional and educational framework quite variable in scope or its ability to create or regulate conservation professionals, so it is worth summarising the most interesting currents.

1.1 Conservation and design

Most valuable architectural conservation theories try to reconcile two key aspects: how to define the original values of the building and how to make them relevant to present and future generations, through a design approach which recognises and composes the complex issues and the operators. Of course, intervention to pre-existing buildings is something that has always happened, in a similar range of approaches to today's (such as the gradual substitution of timber columns with poros ones at the temple of Hera in Olympia; mannerist and Baroque transformations of medieval churches in Rome; or conversion of abbeys into stately homes in Britain).

The key stages presented in this section still inform the debate today and they are discussed as a spectrum of attitudes to building fabric by restorers and designers rather than a historic overview. Conscious efforts to establish an operative attitude towards historic fabric in the form of a conservation theory developed in parallel with the emergence of modern architecture and technology in the nineteenth century, either as a consequence (E. Viollet-le-Duc) or reaction (John Ruskin). The stylistic approach of the former (referred to as *restauro stilistico* in Italian) aims to re-create a formal unity and was expressed in an arbitrary re-introduction of styles and idioms, as in the case of Carcasonne, or the restoration of cathedrals, while he was also an advocate of the prominent use of modern materials such as steel. His philosophy often resulted in extensive new fabric being added, which imitated the original tectonics and made a selective use of contemporary techniques, as was the case in the long restoration of the Olite Palace (Fig. 1.1), destroyed in 1813 (Yárnoz and Yárnoz 1925; Gil Cornet 2004).

J. Ruskin, with his strongly moral, almost obsessive quest for authenticity, influenced a wide range of thinkers and architects (such as the Society for the Protection of Ancient Buildings (SPAB) in England) with an anthropomorphic approach, which sees historic buildings as living organisms with a life and the right to die. Repair, rather than restoration, which often 'freezes' fabric in a worn-out state instead of substituting new, is actively promoted, as in the case of St Giles in Holme (Fig. 1.2; see also Truman (1946)).

These two attitudes often polarised the argument away from the true values of buildings, which are seen as a composition of elements that simply can be either stylistically recomposed or left untouched. The simplistic common origin of both attitudes ignores the technically rich and culturally complex process of design as well as the historic processes buildings are subject to in their life cycle. This historic dimension, on the other hand, can have a disproportionately large influence on various national policies.

When emphasis is strictly on a monument's capacity as a document, conservation will try to be as factual as possible, based on extensive

1.1
A purely stylistic
approach in the
reconstruction of
Palacio Real de
Olite, Spain, by José
and Javier Yárnoz
Larrosa, 1937–66

1.2
St Giles church
in Holme,
outside Newark,
Nottinghamshire,
restored in 1932
by K. Harley-Smith
according to SPAB
principles

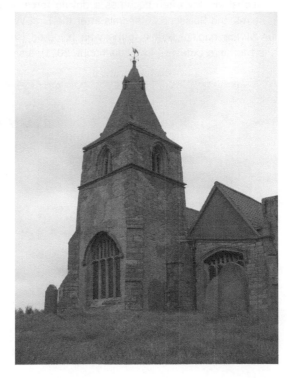

scientific research (archival, archaeological, structural, critical, etc.). The limitation of such an approach (*restauro storico* in Italian) becomes evident when it is highly selective according to criteria dominating the period or the operator's priorities (Fig. 1.3) or when the history of the building is made of complex interactions of phases, not linearly or harmoniously related but often violently transformed. This approach, contemporary to the previous ones, eventually developed into a scientific synthesis (often referred to as *restauro scientifico*), initially by Camillo Boito in 1883, later by Gustavo Giovannoni (1932) and with influences from archaeological methods (such as sequence stratigraphy). Boito's principles (reversibility, distinguishability, etc.) have also set the technical agenda of practice till today.

There is, however, a very critical element that is missing from these theoretical structures, which is the ability of Architecture to generate or synthesise artistic, spatial, cultural and even economic values (Miarelli Mariani 2002). Awareness and attitude to the aesthetic dimension of historic buildings developed alongside the Art History disciplines and the first histories of Architecture, but became more central in conservation following major crises such as the two World Wars or natural catastrophes (e.g. the Florence floods in 1965). The extent and shock of the damage made the need for conservation urgent to everyone and the reconstruction works to be done gave ample opportunity for theories and techniques to be tested, producing a wide range of debate and experience, which is always a positive thing.

National identity often becomes a driving force: in the reconstruction in Reims and Soissons Cathedrals after the First World War the repristination of the original Gothic forms was favoured (see extensive archive at the Memoire Database in Mediatheque 2011), a similar spirit to

1.3
Arbitrary archaeological reconstructions in Knossos, Crete by A. Evans (1911) 'North lustral basin' (K. Theodossopoulos – courtesy ΤΑΠΑ: Hellenic Archaeological Receipts Fund)

that which guided the reconstruction of Frauenkirche later (2005) – see the discussion in Section 6.4 (Fig. 6.37). Camara Santa in Oviedo Cathedral was rebuilt in 1942 after its destruction in 1934 during the Civil War in Spain (Fig. 1.4), using as much of the original fabric, scientifically salvaged, for the crypt (Menéndez Pidal 1960). The reconstruction of Warsaw Cathedral after the city's annihilation in the Second World War, however, used a Hanseatic rather than the pre-war Gothic Revival style, with a very different structural organisation, as it was deemed closer to the spirit the rebirth of Warsaw represented (Warszawa 1939 2011). Apart from the need to repair a trauma rapidly, all these cases have in common the living memory of the original form, often supported by recent surveys, which justified a faithful reproduction not influenced by later cultural values of the ruins.

The artistic emphasis gave rise to the formulations by Cesare Brandi in his *Teoria del Restauro* (1963). His definition has been influential: 'Restoration is the methodological moment in which the work of art is appreciated in its material form and in its historical and aesthetic duality, with a view to transmitting it to the future.' The quest towards restoring the artistic unity permeates essentially the Venice Charter, another influential

1.4
Reconstruction of
the crypt under the
Camara Santa In the
Cathedral of Oviedo
(Menéndez Pidal
1960)

document, but its limitation becomes apparent when considering that buildings are not completed creations of the original designers but they are subject to constant change.

At this point it is important to highlight the difference between the critical acts of conservation and restoration, which refers to recovery of a particular previous state, as well as the difference between the terms in other languages (*restauro* and *conservazione*, for example, have opposite meanings in Italian). The degree and origin of the critical act conditions the approaches in conservation nowadays and paradoxically results from the fusion of the philological research criteria with the innovative spirit of Modernism, when most of the current theories developed. Despite a notorious rejection of history (Bauhaus), the renewed central and intellectual role of the architect in a project and the concept of the unity of the design of outside forms, within the building and the urban context, improved the architectural and technical focus in conservation. The global treatment of historic building is planned and monitored through the Conservation Plan (Burra Charter), an approach which, in some cases, can be unworkable if all technical and non-technical parameters have to be balanced.

A critical approach to conservation is vital as in a sense the architectural values of a building are brought up to date, while respecting its historic and cultural values. The choice and prioritisation of phases, their composition, as well as the removal of those distracting from their reading, becomes then the responsibility of the architect, the informed operator who can perform this critical act (Bonelli 1963). Replicating the original fabric and its tectonics in order to minimise distraction (Fig. 1.5) or relying on new technologies (as in the work of F. Minissi, see Section 6.2) would be left to the architect. Monuments often have complex histories that have to be 'told', so conservation as an approach to present and recompose them always

1.5
San Sisto, Viterbo, an early example of a critical approach to historic architecture damaged by warfare

involves a process of deconstruction to its parts (M. Tafuri) and recomposition (Bellini 1995).

The creative element becomes essential in this approach and is often represented in the work of Carlo Scarpa, especially in Castelvecchio where the fragmented parts are recomposed creatively into new narratives and structural schemes in the composition of medieval fabric and concrete elements. What characterises such approaches is the degree the existing building is used as the basis of a new structural and architectural composition. Completion of an earlier design often offers such possibilities, which become even more critical in designs that have been recently abandoned, as in A. Gaudí's design for Sagrada Familia (Fig. 1.6).

Regardless of the degree of creativity in such a critical approach, the architects or engineers called to design and coordinate have a responsibility that is conditioned by their personal and professional experience, their education and, inevitably, their social attitude. These are in any case the intellectual and organisational skills that are expected from those professions nowadays.

The role of the design professionals becomes increasingly complex as a wider range of vital parameters influences the planning of buildings, such as statutory requirements, environmental considerations, sustainable construction, community engagement. Conservation for public or private use has to include these parameters and the increasing sharing of practice experience has made them specific to conservation, eventually engaging the operators more closely with the monument. Engagement with a monument's values as well as belief in its contemporary relevance are essential in the creative-critical process, becoming the driving force to overcome the technical and statutory difficulties that usually arise. Public pressure can occasionally condition the choices, but architects or engineers are, equally, stakeholders, with the difference that they bring the cultural and technical expertise to make a conservation project relevant to everyone.

1.6
The relatively smooth transition between the earlier neo-Gothic altar and Gaudí's expressionist design in Sagrada Familia (SE corner) is also one between completion of an unfinished work and its creative treatment as the basis of a new composition

1.2 The reintegration of the structural system

Interventions may affect the structure through simple strengthening of the current state, reintegration or addition of new fabric, removal/alteration of the existing structure or treatment as a support for a new one. When these interventions are examined in terms of a conservation theory, the values to be dealt with develop around the integrity and originality of the structural scheme and the tectonics. The structural design could have architectural values that are considered in the composition process, such as the form of a new space roof or the technical resolution of its details, which condition the final effect and the important interfaces.

Similarly to their architectural character, the structural scheme of most pre-modern historic buildings can be very different to current practices. This affects primarily the engineers or surveyors when planning a conservation project as they need specific knowledge about the construction techniques and distribution of loads in schemes they can understand only through experience. Equally important, however, historic buildings are witnesses of the technical knowledge, attitudes and abilities of their creators and their cultures, as highlighted by the new discipline of Construction History (Lorenz 2009), providing a further documentary dimension that needs to be preserved. The latter can be considered as the technical discipline to provide the equivalent of site-specific narratives of architectural history (in this case building technology) that can contextualise the meaning of architecture and place (Bluestone 1999).

Preserving the fabric for the future through strengthening and repairs is usually the basis in every project. In some cases conservation might not even be considered if the structure is deemed to be safe for the new use. This is further supported by the more rational analytical process that determines safety capacity and stability. Therefore, one way to structure the discussion of fabric-related conservation values is through the degree of presence or loss of original fabric and its structural function and character.

Even in technical terms, however, the concept of authenticity can be misunderstood as fabric is linked to a function and performance rather than only documentary values, as even Viollet-le-Duc has somehow acknowledged. Objections to consolidating techniques, replacement or stone cleaning very often stem from a narrow interpretation of ideological criteria in different contexts, as often occurs with the SPAB Manifesto. A more positive attitude would be to clearly identify the original fabric and subsequently understand its strengths and limitations. In most cases, what A. Giuffré was referring to in his texts and projects as 'regola dell'arte' (Carbonara 2010) is exactly a successful performance of the original construction scheme that has been damaged or blurred by ignorant transformations, and therefore its reinstatement may be more than sufficient. Giuffré mentions the case of San Carlo alle Quattro Fontane (Ceradini 1993) where a rigorous analysis showed the high construction quality of the intricate façade, which eventually required very little repair. Authenticity also falls close to the materiality concerns that are discussed in the next section.

Preservation of the original structural scheme or the *regola* can be a purely historicist approach, introducing a dimension of authenticity where there is a need to keep the fabric as part of the historic evidence the building provides. This is where the popular terms minimal intervention, reversibility, distinguishability, etc. find their clearer expression (Hume 1997). This approach is meaningful in the case of important testimonies or techniques that have to be recorded, particularly when the scheme's load-bearing capacity is intact (once this is certified of course). Vaulted structures are often a case in point because of their technical advancements and crucial role in the behaviour of a building (see the reconstructions after the First World War) or its architecture (see the light, yet dated, reconstructions of vaults in San Nicoló Regale, Mazzara del Vallo by F. Minissi or those in St Mary's in Haddington).

Certainly, the minimum disturbance of the structural scheme or introduction of elements compatible in terms of stiffness, durability, movement, etc. is probably the solution that mostly preserves the essence and character of the building. The preservation of the stiffness of a partially removed load-bearing wall in Palazzo Altemps, Rome by Francesco Scoppola and Giorgio Croci (see Section 6.8) or even the acceptance of the original technical errors in the conservation of F. L. Wright's buildings (Section 6.5) are creative approaches in this direction.

There is a fundamental concept that underlines the discussion of theoretical aspects in this section, i.e. the ability to identify a structural scheme that can rationalise the very complex and variable nature of a real structure so that analytical or empirical rules can be used to understand its behaviour (Croci 1987). Deriving its origins from the Kantian *schema*, 'mental images' are a standard construct in engineering science to represent structural actions, mechanical properties of the materials and structural response of elements and behaviour. This becomes an even more fundamental tool to simulate the complex actions, materials and responses of the geometrically, technologically or chronologically diverse historic structures. The discussion that follows is based on the degree of preservation of the historic fabric or system, which is always implicitly represented and conventionally simplified by a schematic abstraction to elements or systems whose behaviour we are confident about (e.g. beams, walls, shells, etc.). Eventually, the bipolarity between historic and current dimensions in a conservation project applies also to the structural scheme.

Starting with ruins or heavily damaged buildings, the extent to which the structural continuity and unity is re-created is a fundamental choice of the designer, as was the case in Koldinghus in Denmark or the Neues Museum in Berlin (see Section 6.4). The extent of the damage determines whether the original fabric becomes the prominent element (partial damage) or is integrated, similarly to *lacunae* in paintings, within a new load-bearing fabric (extensive missing sections). Authenticity is an important issue mainly where fabric has archaeological value, in which case a crumbling structure can be merely strengthened in its current state and a new fully distinguished one can be added that bypasses the weak areas.

Loss and compensation in this case have to be balanced carefully to avoid compromising the visual and structural integrity (Matero 2007). The philological concept of compensation, the efforts to transmit historical or documentary values through reintegration, stands against the incompleteness of form and meaning, which may have been caused by other historical agents (social or political trauma, war) or loss from use, which is not part of the story of the monument. Establishing a balance between loss and gain might be challenged by modern post-structuralist and deconstructionist theories that see visual works as part of a cultural and temporal construct which is constantly changing. Though this contradicts the very essence of structural stability and durability, *sine qua non* for any project, the concept has been integrated in interventions by capturing the moment when a new structure is often brutally juxtaposed or driven through existing fabric (the case of G. Domenig's Dokumentationszentrum in Nuremberg, 2001, or Coop Himmelb(l)au's Dachausbau Falkestrasse in Vienna 1988).

Moving up the scale of the original structural scheme that can be used, modern analytical technology combined with meticulous site and archive research can produce a detailed knowledge of the original structural scheme and secure use of what has load-bearing capacity. This is the case in additions or changes in use, varying from like-to-like replacements (reconstruction of a missing floor, for example) to completely new structures, as will be seen in the case of space roofs over historic atria (Hamburgmuseum or Palacio Cibeles in Madrid in Section 6.1). In the case of modern historic buildings, on the other hand, because of their proximity to contemporary technology, their construction and structural scheme can be considered at the initial design stages of an intervention to ensure that they are straightforward and reliable (Macdonald 2001; Stratton 1997). Conceptually, when no heavy alteration is caused to the fabric, the technical aspects could even be considered as part of the maintenance of the structure, a minimal upgrade rather than a complex updating of the load-bearing scheme.

Adding a completely new structure of an arbitrary stylistic or conceptual resemblance with the original is not an approach that stems from technical considerations. Often the incompatibility between the two systems makes such an approach unfeasible. This was the ground for the rejection of a Gothic Revival reconstruction proposal for Holyrood Abbey Church in Edinburgh by Gillespie Graham and A. W. N. Pugin in 1837.

Technical considerations can also guide the choice of completely bypassing the original structural scheme. There are cases of an active contrast with the critical design process as is usually the case in façadism or extensive interior alterations (Section 6.9). The utilitarian character of the structure, however, needs an analytical or condition survey, which might indicate that the structure cannot hold additional loads and therefore it would be necessary to bypass the original structural system. Both expansions of the Reina Sofia Modern Art Museum in Madrid are steel frames independent from the original hospital, and a more creative approach was followed in the new interior structures that enabled the use of the ruined Koldinghus Castle.

1.7
Upward and lateral
extension of office
block in 80 Mosley
Street, Manchester
using a steel frame
bolted onto the existing
concrete skeleton
(cross-section after
Owen 2007)

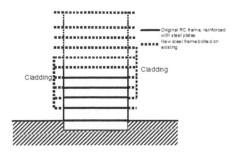

The other extreme of this approach is where the structural scheme is actively ignored and side-lined as found in radical alterations that often lead to the gutting of a building. It is often commercial pressures and inadequate floor plans that make historic office buildings or residences in city centres obsolete and the technical solutions that would preserve a part of the fabric become so complex that they are not feasible any more (see conversions in Section 6.9). The project for the 1921 steel-frame St Catherine's House, London by Rolfe Judd and Buro Happold (Brown 2005) was an opportunity for innovation in the retention of the envelope and the insertion of a compatible and safe new concrete frame. For more recent commercial buildings where the skeleton has a certified load-bearing capacity, the extension can happen upwards, as in the case of 80 Mosley Street in Manchester (Fig. 1.7, after Owen 2007). The building was retained essentially as a framed building, but this was not the case in the radical conversion of the abandoned Eridania sugar factory in Parma to the Paganini Auditorium by Renzo Piano in 1999–2001 (Fig. 1.8). The relation of the performance space with the historic factory, although visually clear, is weak due to gutting all the original floor planes and interior. It has to be said that there is no formula for how much needs to be kept but the intentional destruction of the original character certainly makes the conversion of the building rather pointless.

1.8
Paganini Auditorium,
Parma (courtesy Foto
Studio Carra and
Teatro Regio, Parma)

The critical approach in the conservation of the fabric is expected to be more evident in the purely technical aspects as the building has always to be brought up to date (in terms of regulations, safety, etc.), while being conditioned by its historic and cultural values. This can be simply expressed in a mere technical resolution (strengthening, repair, cleaning, upgrade), while other aspects, such as recomposition, might not be possible due to the nature of the fabric. More complex historic layers with their corresponding structural schemes can even be responsible for the technical problems to be treated, such as conflicting alterations causing incompatibilities, as in the cases of castles updated for military uses or converted to prisons.

Current demands from the built environment are increasingly complex and, apart from the basic need for safety, structural fabric is also linked to well-being, sustainability, etc. Current technical innovation has developed to respond to these demands but also to make almost everything possible. Entire relocations of buildings (Belle Tout Lighthouse away from the eroding edge of Beachy Head in England or the planned move of the Sagrada Familia School – Escolas – in Barcelona) or fragmentary preservation of historic rooms (5 Duke Street, Marks & Spencer, Grafton Street, Dublin) have taken place, and in almost every historic building extra space is provided through a deep basement. There are strong ethical questions about the technology (and the planning permission system), some of which unfortunately can undermine a sympathetic design debate through commercial pressures, and eventually destabilise the architectural unity of any intervention.

Cases of musealisation of original fabric are also observed, often where political or commercial pressures can detach a building from its cultural context. The Arab Baths in Baza (Granada) were completely boxed in situ by a new building by F. Ibañez in 2009, which shelters but has no structural interaction whatsoever with the original building. The remains of the foyer (Kaisersaal) of the 1908 Grand Hotel Esplanade in Berlin were actually moved in 1996 to the new Sony Centre in Potsdamer Platz, with their appearance, enclosed in a glass case, almost highlighting their artificial location (Fig. 1.9).

1.3 Materiality

The architectural qualities of the chosen materials and construction systems are another technical aspect of the character of a historic building that often prevails visually and culturally over the structural aspects. The balance between the historic and artistic characters of the fabric and the need to make them relevant to current culture is more intensive in these tectonic aspects. They range from the visuals of the structure itself and its cultural associations, to the detailing of key elements, the materiality of the envelope or its texture.

Building fabric has been discussed in the previous section in terms of achievement of performance (stability, serviceability), the

1.9
Integration of remains
of the Grand Hotel
Esplanade in the Sony
Centre, Potsdamer
Platz, Berlin

equivalent of artistic value. Technical evidence of the ability of communities, as is visually expressed through the use of materials advanced for their period or increasingly complex systems, highlights those historic values to be transmitted. Most interesting in many cases, but also more difficult to understand, is the emergence of local patterns in the use and expression of materials and geographically bound cultures. Inevitably this study has to start from classifications linked to local construction material resources as in the meticulous work of A. Clifton-Taylor (1972). The more complex interactions with the wider range of resources (not only material, but also territorial and human) eventually require architecture to be analysed almost as an ecological system, as R. Bechmann (1981) attempted in the study of complex works such as the medieval French cathedrals.

The links of fabric with religious beliefs and the local or national culture is a very wide subject and often, considered together with the space and the use of the building, constitutes the character to be preserved. The case of religious buildings in particular highlights many aspects of national cultures towards collective memory and culture. The most stunning example is the Ise Grand Shrine in Japan that probably dates from the sixth century AD. A complete reconstruction of the sanctuary takes place ceremonially every 20 years. A constant renewal of the form in an attempt to keep it eternal guides this culture and cypress beams from the surrounding sacred forest are used every time. A variation of this attitude in ancient Greece saw the gradual substitution of the grooved timber columns of the Hera Temple in Olympia, made from sacred olive trees, with poros stone in Doric style but following the trends fashionable at the time.

The importance of location in the Japanese culture was even stronger in ancient Rome where temples would easily change their plain or archaic forms to follow trendier and larger Hellenistic ones, using more prestigious building materials, as long as they were kept on sacred ground. In other faiths, however, form is tightly linked to ritual. The hierarchically articulated inward-looking space of an Eastern Christian (Byzantine) church is created in a thick load-bearing masonry envelope. Spatial conception has not changed much and even nowadays a church is built as a continuous reinforced concrete surface. This clearly reflects the largely unaltered religious traditions, although no sacred value is attached to the fabric itself. The fifth-century Basilica of St Dimitrios in Thessaloniki, Greece was destroyed during the Great Fire of 1917 but its Byzantine basilical layout was strictly linked to the important shrine of the Saint and was therefore faithfully rebuilt, using modern materials.

Restoration of the materiality is certainly a technical issue as it controls the permeability and durability of the envelope, together with aesthetics. Visual integrity and authenticity are often the immediate aspects to be considered and, if the purely decorative elements are excluded, the issues concentrate on the exposed fabric (load-bearing or framework) or the plasterwork and colour. Historic masonry varies widely in terms of bond, type of units, mortar joints, etc. ranging from the most rudimentary (Dark Ages) to almost sculptural effects (Baroque). The extent to which the original aspects can be restored is conditioned by the damage, and the reintegration of the fabric often follows in association.

All the approaches explained previously apply here and some representative masonry types can illustrate the range of technical problems. Drystone structures often found in prehistoric remains (brochs in Scotland or Mycenaean tholoi in Greece) are cohesionless and depend on friction (Fig. 1.10). The blocks vary from perfectly interlocking formations to pure rubble stabilised by friction, and because of the lack of organic material (no binder or mortar) or historic sources, the only way to establish the original scheme is by critical study of in-situ evidence. The material can be easily dislodged and there is usually lack of historic memory about their

1.10
Rubble masonry in
Dun Telve broch,
Glenelg, Scotland

technique or architecture. The restoration priorities are therefore didactic
(provide an idea of the original impression) and safeguarding.

Ruins of bonded masonry, usually with a rubble core, often suffer from significant loss of binder (abandoned castles) or their original cladding (Roman ruins). Past interventions might have also introduced mechanically and visually diverse elements (Fig. 1.11). The critical point is reached
when the ruin is on the threshold between a recognisable (possibly original)
configuration and an amorphous mass. Original material might also be lying
around but its reuse cannot recover a lost authenticity apart from architectural effects based on the recomposition of the original texture of the surface. Compatibility in terms of strength, stiffness, durability and movement
are fundamental technical criteria for the choice of repairing material and
the extent of reintegration, which is guided mainly by aesthetics (degree to

1.11
Completion of Roman
fabric with modern
brickwork in the
Roman Odeon of
Patras, Greece

which the original fabric is overwhelmed by the new, presence of *lacunae*, architectural composition).

Neoclassical stonework, like the ashlar Georgian masonry in eighteenth-century Edinburgh, has different technical priorities (Fig. 1.12). The thinly joined uniform sandstone blocks have subtle variations in texture and colour, which must be addressed when choosing repair stone compatible in texture, colour, durability, and performance. These are the key criteria, rather than choosing stone from the same quarry, which theoretically is no longer the same material after time has passed (Brandi 1963).

Substituting damaged stone or brick units, however, is not purely a technical issue as is widely thought. Unfortunately conservation is often

1.12
Integration of
new stone into
neoclassical
(Georgian) ashlar
stonework in
Edinburgh

mistaken for this simplistic approach, often resulting from pragmatic and empirical attitudes (Carbonara 1976), and the preservation of neoclassical stonework is very indicative of the consequences. The contrast between new stone and original, uncleaned ones can be very uncomfortable if a project does not study the composition as a whole. This will eventually determine how deep (or not) stone cleaning may have to be and what type of stone exactly goes where. Whichever method is used, the intentions will be good and the results can be a testimony to our technical abilities but they might also undermine the purity of neoclassical lines and their scenic effect on their immediate environment, the essential values of this architecture that the interventions are expected to actualise.

Particular or bespoke details might have to be redesigned and, as with all the areas discussed previously, their location in the restored material-ity of the building must be carefully defined. As modern designs, they would often be viewed as expressive and responsive (Charleson 2005) and this gives a different view of their technical performance that moves from strict, single-purpose functionality towards a more integrative design that addresses a wider range of actions. Typical examples are elements that change behaviour according to the loads they carry, such as ties and their caps.

Integrative design increasingly considers in a more central and quantified manner qualities and actions that traditionally are considered as immaterial. Light and acoustics are often fundamental in an original con-cept (not only in theatres or Gothic cathedrals) and many interventions cen-tre their vision on such aspects, as would happen in a new contemporary design. Shadow play or reflection on material textures becomes a funda-mental architectural consideration in new space roofs or the preservation of archaeological sites. Thermal, daylight or acoustic comfort, however, is demanded anyway by modern regulations for public spaces and if relaxation of the rules is not possible then, as before, a design to moderate or alter these conditions without falsifying or damaging the way the historic building is presented needs to be carefully planned.

1.4 The practice of interventions

The operational aspects that have concerned and influenced many theorists from R. Bonelli (1963) onwards cannot be underestimated. Eventually, the ability to defend and implement a theoretical position and to create a design through the available techniques, historic information or public interests is a professional skill of the same critical nature as the restoration approach itself. The experience, education and range of operators are inevitably vari-able and efficient regulatory frameworks are intended to ensure a minimum quality of service, as will be seen in the next section. The way the profes-sion operates and the empirical approaches that often replace the theoreti-cal sensitivity need, therefore, to be discussed here.

An intervention or conservation project is the result of a broader input by operators nowadays. This is marked by the establishment of the 'design team' in recent decades, which values designers of environmental strategies, for example, in addition to the architects and structural engineers. The roles of the quantity surveyor and project manager also become critical due to the complexity of the projects and funding.

There is, however, an inverse trend in the practice of a project that results from the lack of building skills, such as stonemasons, plasterers, joiners, and even welders, who are experienced in traditional rather than modern techniques. This goes further back to the manufacturing and availability of traditional materials (such as certain types of stone according to specifications, lime mortars or durable hardwoods). Apart from the limitations of carrying out a project to the quality required, the trend highlights a vital lack of continuity in the trades and direct appreciation of traditional systems. When combined with changes in user patterns as in the case of major cathedrals (Taupin 1993), it shows the relationship between the cultural and personal detachment from the historic fabric and its deterioration.

As far as training is concerned, specialist undergraduate degrees exist in heritage administration or linked to archaeology, as well as technical/vocational qualifications. The education of the design professionals is generally only at postgraduate level and learning criteria pertinent to conservation are not specified in UG education, except for some reference to the particular qualities of the existing or historic built environment. In particular for technical matters, in the UK for instance, the guidelines for the training of structural engineers (Joint Board of Moderators 2009) recognise 'understanding that design is a creative process in which experience and a thorough knowledge of historical precedent can inform both intuition and conscious choice' as one 'of the more important design attributes of a competent engineer'. In similar criteria for architects (Architects Registration Board (ARB) 2003) the students are expected to 'demonstrate coherent architectural designs that integrate a knowledge of … architectural histories and theories, of physical, artistic and cultural contexts, and their use in informing the design process', and in Part Two studies an 'ability to devise structural and constructional strategies for a complex building or group of buildings'.

In other European countries such as Spain or Greece a stronger emphasis is given to conservation through specialist courses in architecture, while engineering education addresses the area only indirectly and occasionally with courses on concrete or masonry repairs. In Italy, however, probably due to stronger operative and cultural traditions in conservation, design in existing architecture is promoted as either a specific discipline (*restauro*) in the final year project or a specialist Part Two degree programme.

Dedicated education for design professionals is usually offered in Master and Specialisation degrees. In some countries, such as the UK, an attempt is made to harmonise such degrees at professional level, beyond technical institutions or the university system (Conference on Training in Architectural Conservation (COTAC)). At international level, however, reference is usually made to the International Council on Monuments and Sites (ICOMOS) guidelines that establish standards for training in the conservation of monuments. They can be considered to distil national attitudes and represent the debate of the period (ICOMOS 1993). They attempt to define the wide range of skills conservationists need to develop, and they recognise the need for traditional skills and interdisciplinarity in practice, as well as continuity from university education.

The design practice itself has been largely influenced by early aspects of conservation theory and, in particular, the principles presented by C. Boito as early as 1883 (differentiation of intervention from historic fabric, documentation, etc.). The following axioms apply to a range of situations and are popular among professional engineers (Hume 1997).

- Conserve as found.
- Apply like-for-like repairs.
- Use minimum intervention.
- Replacements should be reversible.
- Replacements and changes should be sympathetic.

Their meaning in engineering practice will be explained in more detail in Chapter 5. They are indicative, however, of the attitude of structural engineers towards their role in intervention projects. The same axioms are somewhat summarised, for example, in the category for Conservation in the British Construction Industry (BCI) Awards which rewards a 'project designed and executed with respect for original fabric, design and form while making a minimum intervention consistent with safety and structural integrity'. This is the way conservation is addressed in mainstream engineering practice, which also highlights a further key aspect, that the project should guarantee good building practice. In line with this, other contemporary aspects should also be considered, such as energy efficiency and Health & Safety planning in the workplace.

The wider spread of conservation means tighter budgets as money must be shared among more monuments: tighter programmes require more control and planning, but offer less flexibility and in-situ inquiry. Conservation, however, demands experience and, equally important, an inquisitive attitude from designers at the planning stage. The management of change is a concept recently introduced in the heritage administration in the UK that aims to align conservation with the fast and intensive changes in the built environment, mainly in planning and city development. The situation highlights the need of flexibility from professionals in the design and management process, and reminds us also of the dangers when the essential control is out of their hands.

Further important tools in the design and construction process are the project contracts, which have developed as a management and operational tool. Current trends, as seen in the NEC3 contracts in the UK, are based on and stimulate good management of the relationship between the contracting parties, as well as the supply chain. These are fundamental aspects in conservation projects where a wide variety of specialist sub-contractors can be involved and the success of a technique depends on very specific traditional materials or bespoke systems.

However, the continuous control stimulated in the NEC3 might require an overarching management structure that is rarely suitable for conservation projects and their scale. Traditional forms of contract should be favoured in general as they give the architect a more central role and an ability to appoint specialist contactors as required; an essential service to any conservation project.

A final aspect that influences modern conservation practice is the availability and sharing of information. Digitisation of public and private archives has revolutionised the access to vital original data about a building, and, together with web-based databases that allow technical and scientific information to be shared, has truly democratised the planning and design process. The increasing presence of manuals in every language, the mobility of professionals and academics, the internationalisation of competitions (mandatory for major projects within the European Union, for example) have allowed best practice to be easily shared and made available. In some cases, information is carefully organised by the local authority into guidance that collects historic details and recommendation of good practice for the use of design professionals or building owners (Giovanetti 1998; Davey *et al.* 1981). The cultural impact on national attitudes and priorities only now starts to become clear and it still needs dialogue within the professional sector to understand its extent.

1.5 Regulatory framework and practice

Moving further into professional aspects and how they condition design practice, it is useful to look at the philosophy and extent of the regulatory framework, especially in countries that actively promote architectural heritage as an important cultural and, even, development asset.

The range in national cultures has been framed by G. Carbonara in his discussion of the contrast between a direct approach that results from the senses (British empiricism) and an intellectual one that results from philosophical assessment (Italian idealism). Without passing judgement at this stage and beyond the attitudes presented in previous sections, such contrasts stem from often 'unshakeable aesthetic positions' (Carbonara 1976; Stanley Price *et al.* 1996). Such national attitudes also result from specific conditions, such as the quality, breadth (Greece, India) and/or depth of heritage (Italy, France), the effect of traumatic national events such as warfare (Greece, Germany, Japan, Poland), the ability to adapt to change (UK) or the quest for modernisation (France), the intensity of a rather recent past and the debate for national identity (USA, Canada, Brazil, Australia), political systems (Russia, China), or the attitude to alien past cultures (Turkey) etc.

Legislation certainly reflects these conditions, and its implementation, as well as promotion of the cultural values of heritage, have been traditionally left to the state. Policy and framework often address, first, the protection of monuments and, second, study, education and development. The most visible aspects are the process of assigning the status of protection to historic buildings and the control on conservation projects. Status designation can be between ancient and historic monuments (UK, France), with further classification of importance (like Grade A and B Listing in Scotland or inscription to the Statutory List of Buildings) or simple assignment of conservation status (Διατηρητέα in Greece, Bene Culturale in Italy). Other aspects of the historic environment that are protected are landscapes in Italy and France, which are strictly linked with the republican values of these countries, or Conservation Areas in the UK.

In terms of policy, the role of government or local authorities varies significantly according to the degree of centralised control. This ranges from quite strong, as in Italy and Greece, where the Ministry of Culture controls directly the policy and its implementation through local period-specific agencies (Soprintendenze and Ephorates), to state agencies (Historic Scotland, English Heritage, CADW in the UK) or regional commissions directly under the local authorities (France, Spain).

The relationships between construction clients, operators and state agencies can have complex facets, depending on the scale of the project or the ownership (public or private). Planning issues depend on the level of national importance, and often special permissions (e.g. Listed Building Consent in the UK) or relaxation of building regulations are required. In legislations with strong operative focus as in Britain, historic environment laws are harmonised with planning regulations (PPS5 – Planning Policy Statement for Historic Environment). Modern uses and the need to guarantee comfortable and safe conditions for visitors increasingly pursue harmonisation with other regulations (Codice dei Beni Culturali 2002 in Italy, for example).

Other aspects of modern legislation address the relationship between central and local government with developers in a more proactive manner, where the role of the private sector and business opportunities is recognised but also regulated (sponsorship or leasing). The structures of financing and tax incentives (such as low VAT in conservation projects) are also viewed as further tools for the development of the historic environment. Non-standard sources, such as the Heritage Lottery Fund in the UK for example, have become vital for the viability of projects with strong community character and participation.

This is linked to the new dimensions of architectural heritage nowadays. Amenity societies were often established as a reaction to destructive attitudes or projects and, today, they function as voluntary organisations that safeguard and promote historic buildings or certain periods, or even attempt to provide a learned opinion or objection to planned projects. SPAB gained status for its ethical principles that were summarised in their famous Manifesto. In the 1960s a stronger wave of societies, such as the Civic Society (1957) or the AHSS (1956) in the UK, sprang up as a protest against the large-scale destruction of heritage. Other societies see themselves as citizens' movements (Monumenta in Greece) or learned voluntary societies

(SIPBC in Italy). There is a stronger public engagement even with the design process that converts existing buildings for community use, and often charity organisations (like Glass-house in the UK) act as enablers for partnerships between local communities and design professionals.

At the other end of the scale lies the significance of certain buildings and sites at international level, usually then designated as monuments. UNESCO is usually credited with the process of listing them and establishing their character and threats. The organisation has no executive or inspectorate powers but relies mainly on the prestige and universal awareness of the importance of these sites. Through its World Heritage Sites network it provides and monitors the criteria that establish the international significance. Edinburgh in the UK is an interesting example of how the various layers of control on the conservation of the historic fabric of the city interact. The Edinburgh World Heritage Trust evolved from the 1970 New Town Conservation Committee into its champion and coordinator of activities through a Management Plan. The City Council has an operative role but also statutory powers (unique in the UK) to ensure owners in buildings of multiple occupation maintain and regularly repair their assets. Finally, Historic Scotland (the Scottish government agency) inspects listed buildings and monitors the measures to safeguard their character.

The role of non-governmental societies at this level (such as ICOMOS or DoCoMoMo) is consultative and aims to broaden awareness of global issues. They promote collaboration and share experience in conservation, at both technical and scientific levels, which is very helpful for countries in the developing world or in the cases where further pressure is required to protect monuments of international importance.

International charters were established early in the twentieth century in recognition of the need for international collaboration (and probably in line with other international initiatives such as CIAM). Very often they are much better known and used among professionals and organisations rather than well-developed theories or national regulations. The Athens, Venice and Burra Charters are the best-known ones and they mostly reflect the attitudes of the times in which they were written. They are distinguished by their approach to what constitutes the key value in a historic building or site, ranging from the artistic (Venice) to the more cultural (Burra), and the level of operations they promote (the Conservation Plan in Burra).

While such initiatives are always positive, they should not be over-emphasised as they are often private initiatives, based on good will, that have not been the result of national, scientific or technical debate among the respective communities. At the same time, problems are likely when cultural values become too eminent or conservation projects aim primarily to bring historic works up to date. UK policies put too strong an emphasis on public fruition which removes the distance that is often associated/needed in the appreciation of history. This makes monuments subject to budget changes and priorities, promoting or sidelining aspects that are not fashionable, politically 'correct' or even commercial enough.

Chapter 2

Structural theory in historic buildings

When preparing an intervention that respects and integrates the character of the historic building, the structural scheme of the building must be defined. A desktop study and in-situ assessment will determine its present condition but also the technical characteristics and historic importance of the fabric. This process is aided by knowledge of how the building was planned and its members sized, in other words, the methods of structural design and how closely the effect of the loads was resisted by the structure.

Subsequently, regarding the planning of conservation and the intervention, it is important to understand the capabilities and limitations of the present analytical and sizing methods. The second part of this chapter deals with how these methods can be chosen and applied. The response of the main structural systems, in a very broad classification for the scope of this overview, will be discussed in more detail in the following chapter.

Most of pre-modern historic structures did not use scientific structural engineering theory consistently as a design or assessment tool. The few exceptions are testimonies of an attempt to formulate and transmit empirical rules, as in the case of the R. Gil de Hontañón (Sanabria 1982), or to balance contemporary mathematical theories with engineering sense (Poleni's (1748) report on St Peter's dome in Rome).

The topic is, of course, quite broad and the reader is guided to more comprehensive overviews of the development of structural theory and analytical methods, such as those written by Edoardo Benvenuto (1990), Jacques Heyman (1995), R. J. Mainstone (1998), or papers presented at the Construction History or Structural Analysis of Historic Constructions specialist conferences.

2.1 Early engineering assessment

One thing that has always to be remembered is that in the past knowledge of structural behaviour or even good building practice was disseminated by completely different means than today. Education, social structures, the lack of paper and the very restricted circulation of books, combined with the nature of the construction industry (based on skilful masons and their apprentices) made written records for the transmission of knowledge difficult or even pointless.

Several alternative systems have been proposed. They are rarely useful in providing exact knowledge of an original design or its performance, but they serve the purpose of helping us to understand the act of overcoming material and social constraints through creativity and innovation. The prevailing theories believe that masons were organised in a tightly knit and hierarchical system of crews and they would learn in an instinctive manner directly from previous buildings.

Based on the available evidence from archive material, it is most probable that the original masons in the course of the High Gothic period had not established an independent set of structural design tools, but that they had intuitively developed some rules of dimensioning individual members, incorporating them into spatial proportioning methods. Even when Mignot, for example, asserted during the debates on the design of the new Milan Cathedral in 1400 that 'Ars sine scientia nihil est' (There can be no technology without science), he was referring to a 'geometric science', expressed in a predetermined geometric grid for the whole building (Mainstone 1998). These proportioning methods would then be transmitted either orally through the corporations of the masons or, more effectively, by using the building itself as a compendium of solutions to specific structural problems (see more on Gothic design in Acland (1972), Kidson (1986), Fitchen (1961) and Huerta (2006).

Other methods on the preparation and structural assessment of primary elements, and transmission of their good practice may have included models, but most of the surviving ones were intended for presentations, like Wren's or Michelangelo's, with no indication of structural exploration. Tracing on sand and stone slabs was more efficient and very often included proportional studies, as in the case of the Hellenistic Temple of Apollo in Didyma (at the adyton), the tracing floors (or épures) in the Gothic cathedrals of York (Mason's Loft) and Wells (Holton 2006), or the arch construction study in Roslyn Chapel, outside Edinburgh.

Regarding methods of sizing members, the earliest evidence available of any attempt is from the Middle Ages. Manuals, such as Villard de Honnecourt's sketchbook (Carnet) or German and French palimpsests (e.g. for Reims and Strasbourg cathedrals), make no attempt at providing construction rules and it is only after the first Renaissance treatises (Alberti, Serlio, Dietterlin, Piranesi) that the first construction manuals emerged. Rodrigo Gil de Hontañón (1500–1577) was a Late Gothic architect and mason who balanced the declining Gothic style, where the construction process is more evident, and the revival of Classicism that promoted increasingly aesthetic criteria. He attempted to produce a complete theory, publishing seven formulae that evaluate the necessary depth of a buttress to support rib vaults or arches (Sanabria 1982). These are divided into arithmetic or Structural Rules, where closed expressions are given, and Geometric Formulae, which derived from geometric constructions.

The very interesting Third Structural Rule assesses the depth of an outer wall buttress for a hall church as double the width W which is:

$$W = \tfrac{1}{3} \cdot \sqrt{H + \tfrac{2}{3} \cdot \sum R}$$

where H is the height of the buttress to the springing and ΣR is the sum of the perimeters of all the ribs converging to the buttress, measured from the springing to their keystones, usually a quadrant. Clearly, the units are handled inconsistently and this is a result of their purely empirical nature.

The second set of rules, however, establishes also the height of the (spandrel) wall of an arch or barrel vault, apart from the buttress depth. The First and Second Geometric Formulae give a height of $2.081R$ and $2.894R$, respectively, where R is the radius of the arch profile. Sanabria (1982) regards the two methods as if two experiments were conducted in which two arches of a given span and slightly different buttress depth were loaded until they failed. Although the concept of dimensioning individual structural members was just emerging at that period, it is very possible that Rodrigo Gil had the chance to identify and test some samples, as he was the architect and director of several construction sites.

The scientific process that Gil espoused found clearer expressions in later times as the science of mechanics developed. It is not clear whether Christopher Wren performed a mathematical sizing of his dome for St Paul's in London (1675–1709) with the aid of Robert Hooke, the man who had proposed the relation between forces and the deformation they cause. However, the first application of mathematical rules occurred when G. Poleni challenged the opinion of three French mathematicians invited in 1742 to give their opinion about the structural instability of the dome of St Peter's in Rome (Poleni 1748). Following a rational process that combined experience with the use of the catenary concept, he assessed the dome's strength (Fig. 2.1) and eventually applied a series of chains that were experimentally evaluated in his laboratory in Padua.

Couplet (1730) and Coulomb (1776) established the 4-bar chain collapse mode for arches and Méry and Moseley formulated the theory of Lines of Thrust between 1830 and 1840. Rondelet published the first set of practical dimensioning rules in 1810 after he conducted experiments on arches and vaults. These rules remained in wide use until Ungewitter and Mohrmann (1890) and Körner published new and more accurate tables in 1890 on the calculation of the safe weight V_0 and corresponding thrust H_0 based on the assessment of the line of thrust. These values and the unit weight applied as a surface load d can be found in the table reproduced by Heyman (1977) (see Table 2.1), for various unit materials and height to span (f/s) ratios. The forces were calculated for a unit plan area and they included allowances for the ribs.

The essence of the design of structures in pre-modern history is that they were often oversized, with a large degree of redundancy which is beneficial today as it usually means that the structures can be overloaded if an addition or change in stiffness occurs. Also, historic structures often

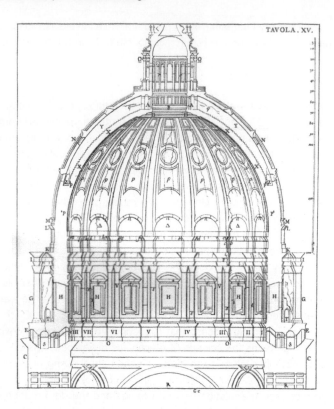

TAVOLA. XV.

2.1
The crack pattern of St Peter's in Rome from G. Poleni's seminal structural assessment (1748)

Table 2.1 Thrust H_o, safe weight V_o and unit weight as surface load d for quadripartite vaults for a unit plan area (KN/m²) (Ungewitter and Mohrmann 1890)

Height/span (f/s)	1:3		1:2		2:3		5:6 to 1:1		
	V_0	H_0	V_0	H_0	V_0	H_0	V_0	H_0	d
½ lightweight brick	2.3	1.6–1.8	2.6	1.1–1.2	2.9	0.9–1.0	3.4	0.8–0.9	1.6
½ strong brick	3.1	2.2–2.4	3.5	1.4–1.6	3.8	1.1–1.3	4.5	1.0–1.1	2.3
¾ strong brick 200#mm	4.2	3.0–3.3	4.8	1.9–2.2	5.3	1.6–1.8	6.5	1.5–1.6	3.1
sandstone 300#mm	5.7	4.2–4.5	7.0	2.8–3.2	7.5	2.2–2.5	9.0	2.1–2.3	4.0
rubble	10.0	7.1–7.5	12.0	4.8–5.5	13.0	4.0–4.3	15.0	3.5–3.7	7.2
Lever arm h/f		0.85–0.75		0.8–0.7		0.8–0.72		0.8–0.75	

have an intuitive regular and even symmetrical arrangement of strength and stiffness (both in plan and elevation). This is the reason why emblematic structures such as classical Greek temples (Parthenon, Apollo Epikourios) or Gothic cathedrals (Soissons, Reims) survived well until affected by direct human action (often it took bombardment to partially destroy them).

Lessons learned from failures (collapse of Beauvais Cathedral in 1284, major fires in London or Istanbul) caused dramatic changes in practice, ambitions or regulations. The usual cycle of conservatism and risk-taking was longer than today and the two collapses in the very slender choir of Beauvais marked clearly the end of structural experimentation in Gothic times. It is usually, however, the emblematic buildings that have survived and it is difficult to know how far technical advancements might have influenced the poorer dwellings of people around them, such as the houses in ancient Athens built probably in adobe.

2.2 Modern historic buildings

As seen in the previous chapter, modern structural engineering theory and practice, similar to conservation theoretical formulations, developed alongside the modern materials, systems, manufacturing and construction processes from the eighteenth century. The analytical tools became more advanced and have often been established in a mathematics and physics framework: theory of elasticity, plasticity, failure criteria, strength of materials. Other advancements in the industrial era are a scientific knowledge of strength, while since the 1950s the design process is based on stress limits, the application of safety factors and the concept of risk, and the definition of limit states. A recent trend that addresses structures that cannot be simulated or studied theoretically with much certainty, such as fire response, is performance-based design. Although a process based on careful experimentation, in-situ studies and statistics, it is an interesting approach to the empirical design process in pre-industrial times.

Modern practice has used materials and components as systems, and theory, education and regulations have become more standardised with the trend to harmonise around zones with common economic interests, such as the Eurocodes in Europe. There is once again a similarity with medieval times, when the mobility of stone masons beyond state boundaries, politics or conflicts allowed the dissemination of good stoneworking practice and the establishment of certain typologies.

These facts are common practice nowadays but have also a fundamental relevance for some early modern structures such as reinforced concrete and steel. Concrete was sized and built following certain systems (Hennebique, Coignet, Considere) while the modularity of steel permitted standardisation of analysis. The confidence in these new systems and the quality of the manufacturing on an industrial scale also produced further mixed types where the benefits and efficiency of each system were combined favourably, as in the textile mills of the time. The skeleton would be cast-iron columns, with wrought-iron beams and concrete or brickwork jack-arches or timber decking, as occurred often in industrial buildings such as textile mills (Fig. 2.2).

The results are more efficiently designed buildings but with less redundancy and degree of adaptation. However, they fit more closely the current process of analysis and the effects of any interventions can be more carefully evaluated, as is the case of new roofs over atria in nineteenth-century museums (Hamburgmuseum, British Museum).

2.2
Cast-iron column
and timber deck in
a nineteenth-century
textile mill
in Nottingham

Treatises and manuals also became more specific to the construction industry. Publications with detailed illustrations of construction details and processes, including empirical rules for the choice and sizing of elements, became widely available in every Western industrialised country. These are a valuable source in the understanding of certain details and the rationale behind them, especially those not practised any more by the construction trades (ornamental stone carving, production of lime mortars, wrought iron manufacturing, historic glass, riveting, etc.). A selection of such manuals (Nicholson 1823, 1828; Burn 1873; Campin 1883; Nacente 1890; Barberot 1895; Rovira 1897; Newlands 1900; Donghi 1906; Adams 1906; Purchase 1929) can be found in the references.

2.3 Structural engineering in historic, pre-modern buildings

Modern analytical and design tools are based on different assumptions and address a much more intensive design and construction process. If they are carefully interpreted and adapted in the context of historic buildings that need conservation or are about to receive a major intervention, they can give information on a wider variety of responses and allow more control on, especially, very delicate operations or maintenance strategies.

When the load actions are determined, a wider range might need to be considered, especially for public buildings that have to perform many functions. Some actions, such as accidental or high imposed loads or thermal expansion, are usually accommodated by high robustness. The effect of others, such as fire action or impact, although implicitly considered, might have to be more precisely determined. Masonry itself might not lose its integrity in an intensive fire (apart from surface damage) but a damaged floor in a building will eventually cause the collapse of the load-bearing walls, either by impact or instability once the diaphragm action is compromised.

Traditional methods of construction, unfortunately, utilise many materials that can provide the fuel for a fire. The necessary prevention involves fire detection and suppression by sprinklers that might be intrusive in the aesthetics of an enclosed space. Equally effective strategies prevent the fire from spreading or reaching the vulnerable elements. Compartmentation with automatic fire doors (new or reconfigured existing ones) or ventilation arrangements is a less intrusive and often effective strategy, which depends on careful planning and management. Fire-proofing should be applied in a way that disturbs the fabric to a minimum (for example, between the floor joists). Overall, fire safety engineering is the most adequate preventive action but in most cases a regulations relaxation from the authorities might be required (such as Listed Building Consent in the UK).

An action that has gained attention by public clients and designers lately is blast resistance against terrorist attacks. The Scottish Parliament in Edinburgh is one of the first to be designed in this manner and this affected the conversion of the Grade A-Listed Queensberry House, considered the finest townhouse in Scotland at the end of the seventeenth

2.3
The gutted interiors of Penicuik House, Midlothian, UK (courtesy of the Scottish Lime Centre)

century, and now part of the parliament complex (Lewis *et al.* 2006; González-Longo 2008). In this case, protection concentrated on mitigating hazard from fragments of shattered glass from the windows hitting the users. Laminated glass was mounted on 'slam-shut' windows, in which the casement opening outwards would be forced to shut by the blast. An elegant solution was also devised for Queensberry House to locate these windows inside the original ones, creating new contemporary interiors without disturbing the original seventeenth-century masonry walls and plain lime-washed exterior.

Apart from these specific actions, long-term effects will always have to be considered in any new fabric because of the long life that is expected from the historic building it is added to. The original materials might not be of standard or certified production, but if like-for-like are applied, as is the correct practice, then coefficients of safety do not have to be applied as the structure is already there. Whenever assessment of the safety factors, material properties or current load-bearing capacity of the structure is necessary it will have to be done with in-situ tests or the methods that will be shown later in Chapter 4. It has to be mentioned here, however, that the cost of assessing these properties will have to be balanced with the costs and time since, especially with the safety margins, a minimum reduction may result which is not worth the effort (Beckmann and Bowles 2004).

The definition of imposed and other loads that are suitable for the existing structure can either come from current regulations (such as Eurocode 1 or BS6399) or from the building itself. In-situ tests, as occurs in masonry arch bridges, are not always possible so critical observation of the current state combined with desktop study of original or archive project material, photos, witnesses, etc. might provide valuable information.

So far as stress conditions are concerned, the usual range applies (uniaxial stress, biaxial bending, shear, torsion) but their composite action is often complicated by uncertainties or large changes in the structure. The problem can be bypassed if the analysis concentrates on the possible collapse mode rather than the distribution of the loads, what is usually classified as limit-state analysis against elastic or continuum analysis, as will be seen in the next sections. This debate often regards materials that are brittle or have cracked, like masonry, stonework and concrete, while elastic analysis may be directly suitable for steel and timber, if they are not in a state of excessive deformation.

2.4 Analysis methods

Numerical methods that are based on elasticity or the assumption of a solid continuum are not, of course, suitable to the nature of masonry, the most typical material. Such methods, however, either as standard calculations or Finite Element (FE) modelling, can yield useful information such as the distribution of loads (Fig. 2.4), deformations or potential failures. They can be helpful in the design of a repair or when new structural elements have to be sized and located. Combined with the ability of FE models to represent

2.4

Elastic analysis of cross vaults from the nave of Durham Cathedral using an FE model with orthotropic shell elements: transverse stresses at the extrados (kN/m²)

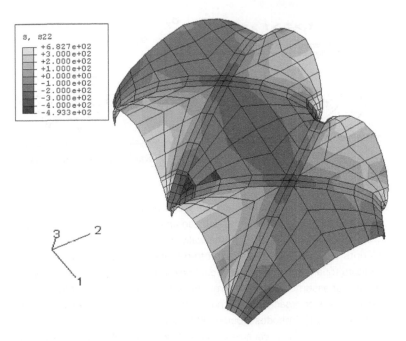

```
S, S22
  +6.827e+02
  +3.000e+02
  +2.000e+02
  +1.000e+02
  +0.000e+00
  -1.000e+02
  -2.000e+02
  -3.000e+02
  -4.000e+02
  -4.933e+02
```

complex geometries such as vaults or roof trusses, such mathematical models are very attractive in practice and research.

Attempts are constantly made to adapt the continuous elastic domains to the nature of masonry. The key problems result from the discontinuity of the material, due to different materials or the presence of cracks and voids. Simulation as orthotropic material simplifies the perceived anisotropic behaviour but a further complexity is in the failure criteria that have to combine different strengths and responses in each orthogonal direction. In the case of dry construction (classical temples or drystone walls) or heavily anisotropic materials (ruins in random rubble) the analysis becomes even more complex numerically. As will be seen in the next chapter, the brittle nature of masonry (no compressive residual strength after a crack and tensile strength can be negligible) sometimes is helpful as it avoids the need to simulate the post-yield plastic behaviour.

Current work in analytical research attempts an increasingly complex coupling of orthotropic elastic behaviour with anisotropic plastic behaviour, aiming to establish a general failure criterion and using macro-modelling approaches that explore the generation of failure through Continuum Damage Mechanics. In this area, P. B. Lourenço and colleagues (Lourenço *et al.* 1998; Lourenço 2005) defined an anisotropic yield criterion for the behaviour of masonry in-plane stress by combining criteria for isotropic quasi-brittle materials including two new components based on the Hill failure criteria (for ductile materials in compression) and the Rankine criterion (for brittle materials under tension loads) (Fig. 2.5). A good overview of relevant

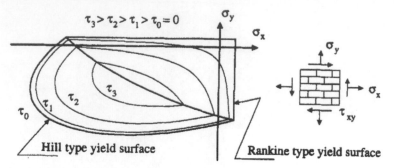

2.5
Composite yield-line
surface with iso-shear
stress line (Lourenço
et al. 1998)

analytical methods and their philosophy and applications can be found in Lourenço (2002) and Atamturktur and Laman (2010). The complex post-failure behaviour was supported numerically by modern algorithmic plasticity concepts (implicit Euler backward return mapping and consistent tangent operators for all regimes). For less complex structural regimes, even biaxial criteria that assume no tensile strength and can be implemented with the smeared crack approach into FE models have shown to give reliable results (Sinha et al. 1997; Theodossopoulos et al. 2003).

Other analytical areas such as micro-modelling or discretisation simulate the failure that is expected to occur in the joint rather than the unit, in particular its interface with the unit. Such modelling requires complex tests, even more complicated for out-of-plane response including buckling. High computational load and the need of specialist experience mean this technique is not widespread among practitioners but, when combined with homogenisation's practical insight, intermediate meso-modelling techniques can map information from the behaviour of an element analysed in micro-scale to the continuum in macro-scale (Milani and Lourenço 2010).

The application of such approaches has gradually enabled the analysis of more complex structures (Roca et al. 2010), showing a growing confidence that modern analytical tools can handle large-scale problems under a wide range of actions: Saint Mark's Basilica in Venice, several hall-type northern Spanish churches, behaviour of urban settings under seismic loads, the strengthening of the foundations of the Pisa Tower, etc. The difficulties in modelling entire structures increasingly interest practice, where the geometric and material complexities become apparent, but most significant programmes are applied to mainly singular monuments, such as the Daphni monastery (Miltiadou-Fezans 2008). Larger churches are represented by their cross-section or a critical part (Clemente et al. 2006; Theodossopoulos 2004) and entire structures are modelled for their seismic response, such as Fossanova (De Matteis and Mazzolani 2010) or Agia Triada in Drakotrypa (Manos et al. 2008).

The capabilities of numerical analysis are, however, conditioned by information from the monument about material properties, the geometry or failure as well as their validation by experimental programmes (to study issues like eccentrically applied loads, durability, 3D structural behaviour, etc.).

Discrete element methods (DEM) may also be suitable in discontinuous structures that lack cohesion, such as prehistoric drystone buildings or thin-jointed stone walls and arches that have lost their bond. This category of theory is based on assumptions that the blocks are kept in contact by gravity and friction. The method can be quite complex for the simulation of 3D performance because of the rotational components and interactions between the friction planes and contact laws, therefore 2D simulations could yield more useful results on the stress distribution and collapse modes. Regarding the assessment of dynamic stability, rigid-body motion analysis can produce some meaningful simulations of the rocking of museum objects or even the stability of classical temples (Ulm and Piau 1993). Critical elements, such as columns or vaults, can be studied both individually and within the boundary conditions of their structure. Most columns in such temples are composed of short drums and their DEM analysis assumes rigid blocks and frictional joints. The analysis, for example, of the Pronaos of the Parthenon (Psycharis *et al.* 2003) under seismic loads showed the adverse influence of drum imperfections on the stability of that area of the temple.

Later chapters and the discussion of case studies will show how these methods are used in interventions. For example, the FE model of the nave vaults in St Francis' Basilica in Assisi, following the damage from two earthquakes in 1997, assisted the restorers to understand the collapse mode and degree of safety of the nave as well as to plan and size the fibre reinforced external ribs that strengthened the vaults. Similar use was made of FE models prepared for Beauvais Cathedral (Coste 1995).

2.5 Limit states

Assessment of the stability of a structure made of rigid blocks, or large discontinuities through the definition of its limit state, can also be carried out through the location of the thrust line within the structure. The method usually applies to arches and stems from the limitations of elastic analysis to locate the many possible equilibrium states in masonry structures, which makes it meaningful to consider them only at their ultimate state, before collapse (Heyman 1995). Kooharian (1953) and later Heyman (1966) made the first applications of the plastic design philosophy to voussoir arches, according to the following principles:

1   Masonry has no tensile strength.
2   Stresses are so low that masonry has effectively an unlimited compressive strength.
3   Sliding failure does not occur.

Application of thrust line analysis (Heyman 1995) was used by many researchers to understand the performance of historic arches and vaults in a way that can be assimilated by the design community. A long-term study by S. Huerta (2006) of Gothic structural rules showed they were used by stone masons as a method to register stable forms and transmit knowledge. On a larger scale, the equilibrium of a cross-section of a Gothic church, such as Exeter Cathedral, can be evaluated through a more comprehensive limit-state analysis that includes contributions to the location of the thrust line by elements of the structure such as the pinnacles (Maunder 1995). J. Ochsendorf defined failure mechanisms of circular arches caused by spreading supports (Ochsendorf 2006). The parametric study related the span increase with the voussoir thickness and showed that: a given increase may be accommodated by various equilibrium configurations; some arches may exert a five-fold thrust at collapse compared to the initial minimum one; the theoretical increase to cause collapse is almost linear in relation to the thickness ratio.

The application of the method will be shown in a cross vault from the aisle of Holyrood Abbey Church in Edinburgh (Fig. 2.6), a partially collapsed Gothic church, that was studied experimentally and analytically. Considering the vault sliced into a series of parallel rings is a safe, statically admissible solution, since experience and tests show that cracks usually develop parallel to the wall and the nave arch.

Once that portion of the vault is detached, the rotational degree of freedom (DOF) parallel to the crack is released. The remaining web is weakened due to this hinge as the resulting moment redistribution causes rupture of further sections to progress more rapidly. The fracture (discontinuity) lines are considered to form along the intersection of the webs (groins) and the individual strips are supported on the ribs.

To study the ultimate limit state of the cross vault model, each strip was examined separately and a crack pattern was considered, representing the lower limit of each arch. Cracks were assumed to form at the crown (intrados) and the haunches (extrados), where the thrust line was tan-

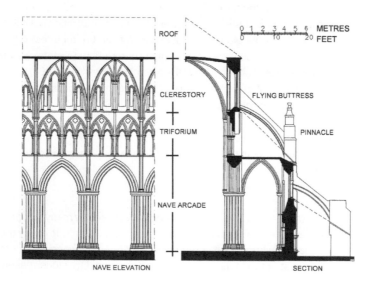

2.6
Reconstruction of the nave of Holyrood Abbey Church. The intact parts are the South aisle (on the right) and the walls up to the springing of the flying buttresses (drawing by Cristina González-Longo).

gent to the surface of the arch. The generic case of a pointed arch defined by a crown angle ϕ_0 and a springing angle ϕ_1 was studied. The minimum required thickness of the shell that is sufficient to contain the line of thrust was evaluated by assuming the hinge at the haunches forms at an angle ϕ from the vertical. At that point, the thrust is offset by half the thickness, t.

The minimum required thickness of each web can be assessed at each slice defined by the springing angle ϕ_1 (Fig. 2.9).

The transverse web is safe as the thickness of the shell (35 mm) is always above the minimum thickness required, which can reach a maximum value of 3.6 per cent of the radius, or 34 mm. The sharp change of sign after $\phi_1 = 62°$ indicates a reverse in the face of crack. For $\phi_1 = 90°$ the minimum thickness required exceeds the actual geometry, but actually

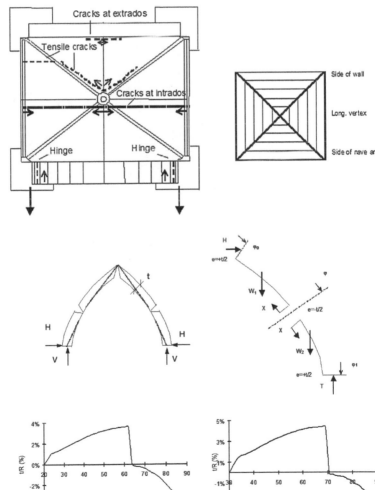

2.7
The failure pattern in an aisle vault in Holyrood Abbey church after movement of front supports and fragmentation of the vault in individual pointed arches for limit-state analysis

2.8
Evaluation of thrust exerted by a pointed arch and minimum required thickness. Location of thrust line and cracks; Free Body Diagram and definition of geometry.

2.9
Minimum required thickness of the sliced webs of the vault. (a) Transverse web: $\phi_0 = 20°$ and $\phi_1 = 85°$; (b) Longitudinal web: $\phi_0 = 30°$ and $\phi_1 = 90°$

$\phi_1 = 85°$ so the arch is overall safe. The study of the longitudinal web showed that the stripes defined between $\phi_1 = 60°$ and $70°$ required a thickness of 42 mm (4.4% of R), after which point the face of the crack was inverted. In that area, cracks are expected to form at an angle $\phi = 42°$ or $12°$ from the crown ($\phi_0 = 30°$). The continuity of the web in the longitudinal direction and the boundary conditions along the groin reduce significantly the deformations during the response of the vault to dead load. Disturbance of these conditions due to movement of the abutments are therefore critical for the safety of the web.

Chapter 3

Materials, building types and their failure

Fabric in historic buildings is characterised by materials that nowadays are not produced or used any more in the traditional and historic manners of the past. In many cases there has been progress from historic practices due to modernisation of the industry, better knowledge of the materials' properties, more efficient construction processes, etc. This applies to brick-work and even stonework, which in modern buildings assume a tectonic character in the expression of the envelope distinct from the purely load-bearing function of traditional masonry (Dernie 2003). Concrete and steel have also evolved from their first applications in the nineteenth century into more durable and strong systems that have expanded the capabilities of the frames and shell forms.

Such differences are even more evident in structural typologies that directly result from the mechanical properties of materials or specific social, cultural and technological conditions. Such structures are often con-sidered as pre-industrial and only shell types have found direct applications today, while any other form directly inspired by abandoned past types is often classified as 'innovative' with a limited effect on modern structures.

Another crucial difference from historic fabric is the gradual detach-ment of the envelope from the load-bearing elements, which become a skel-eton. The modernist structures that highlighted this trend by the implementa-tion of the curtain wall since the 1920s have also a historic value for being witnesses of this technological evolution and they are also discussed here.

3.1 Main structural materials

It is important to briefly review the key historic materials and fabrication proc-esses before their structural performance and pathology in building forms. For each type, the key mechanical properties and responses to loading are outlined, focusing more the discussion on analytical methods from the pre-vious chapter. More comprehensive and broader historical overviews of construction systems have been published for various, more specific areas, such as: twentieth-century developments or concrete by James Sutherland (Sutherland 2000; Sutherland et al. 2001); traditional materials in England by

Alec Clifton-Taylor (the reviewer of geographic material culture in Pevsner's Buildings of England series) (1972); Roman construction by Cairoli Fulvio Giuliani (1992); iron-framed textile mills by Tom Swailes (1998); timber by David Yeomans (1985, 1999); Byzantine structures by Robert Ousterhout (2008); etc.

3.1.1 Brickwork and clay masonry

Masons would always intuitively consider the high compressive strength of stone and brick and employ thick lintels or very strong stone only in areas of high bending (doors, windows). Alternatively, vaulted arrangements could profit from compressive strength if the masons could understand axial loads and form their structures accordingly. This process was often developed through an intuitive choice of catenary forms, which became more conscious and rational in modern times.

Brick has existed in a wide combination of clay-based materials with admixtures, kiln fired or not. It is traditionally considered as a low-tech material strictly related to local resources so instead of progressive historic development, bricks should, rather, be discussed within such conditions.

Adobe blocks are the most ancient types and are still used today in traditional (i.e. less industrialised) construction or new applications in sustainable systems for their life-cycle efficiency. Major-scale historic applications are still preserved in the walls and barrel vaults of the Ramesseum in Egypt (1277 BC), and a technique of false vaulting still survives today. Low-durability mud bricks were also commomly used in city dwellings in ancient Greece or the Middle Ages, leaving very little trace, in contrast to the great and more durable monuments of their times that they were built around.

The base materials are easily sourced clay (mud) and sand, and the mixture is moulded into blocks and then left to dry and harden in the sun. The strength is much lower than fired bricks and straw or even horse hair is used as reinforcement. They are not suitable in wet or rapidly changing climatic conditions and are usually well preserved in hot and dry areas. Their conservation must ensure that moisture is driven out and no incompatible or too stiff materials are added. These problems appeared in the mud-brick walls of Capo Soprano in Gela, Sicily, when restored by Franco Minissi in the 1960s by clamping them with tempered glass panels. Although intended as a subtle showcase, the panels soon created a greenhouse effect damaging further the fragile remains that had been conserved in sand.

Fired bricks demand more energy resources that were not available everywhere until industrial processes and fuel became widely accessible. They have comparable strength to stone, but they can be fabricated in units that are easy to handle. Their properties can be controlled over large batches and due to their porosity the bond with mortar can be stronger. As with modern construction, strength, construction quality and modularity were the key characteristics throughout history.

Roman structures are examples of highly specialised construction in pre-industrial times that made a versatile use of the material. A series of general-purpose square tiles was available (*bessales* of side ⅔ of a Roman foot or about 20 cm, *sequipedales* 1.5 feet or 44.5 cm, and *bipedales* 2 feet or 59 cm wide) but a very wide range of bespoke units was also fabricated. Since many buildings were plastered or clad with high-quality stones, the masonry was not exposed and therefore could be rougher, so the bricks were often split into smaller units (Fig. 3.1). Brickwork (*opus testaceum*) was made of bricks bonded with very strong pozzolanic mortars in thick joints creating durable walls that can carry high amounts of load even later when other structures were added on them.

As for the equally important modularity, it became more crucial in the case of arches, when entire *bipedales* were used, or exposed brickwork patterns in late antiquity. The sheer number of bricks produced in the Roman period influenced heavily the construction patterns until the Middle Ages (in Rome no new bricks were manufactured until the sixteenth century). The size and durability of Roman buildings became a model for later masons and they served as paradigms for major church buildings in the Romanesque period.

The Roman brick construction practices affected even the planning of masonry buildings until the Renaissance as often reference was made to the systems followed widely by the Romans and interpretations of Vitruvius' treatise. In principle, the wall was made of two leaves in coursed brickwork filled with a conglomerate in Roman concrete (known also as

3.1
Core brickwork in Caracalla Baths, Rome (with kind permission of the Ministero dei Beni e Attivitá Culturali – Soprintendenza Speciale per i Beni Archeologici di Roma)

opus caementicium). The variations were not so much in the bond (where bricks were mainly used) but in the combination with stone blocks of tufa of similar size: in *opus vittatum* the courses almost alternate (a technique that also continued in Byzantine times) while in *opus mixtum* bands of brickwork would frame a diagonal grid of tufa blocks (referred to as *opus reticulatum*).

As mentioned, brickwork is related to industrial production and it became a characteristic building material of cultures advanced in the field. Where bricks were less feasible, the masonry would be mixed as in the case of the Byzantine world or even large housing blocks throughout Europe until the nineteenth century. A variety of building materials would be used in thick mortar joints and when exposed an attempt was made to create patterns, like the cloisonné masonry in Byzantine churches (especially from the ninth century onwards).

Exposed brickwork often characterised countries strong in industry, such as the UK (Victorian textile mills or housing in England) or commerce (Hanseatic architecture in the Baltic) or where clay material was abundant (central Spain, North Italy, the Netherlands). Although dimensions would be largely based on the builder's ability to handle the unit, regional differences would be in the colour (depending on the quality of the local clay or admixtures and the firing process) and the patterns (for example, Dutch bond). Bonds and modularity characterise these more recent types of historic brickwork and solid walls were made where the courses are well overlapped along the bed joints or the thickness. Some successful joints are formed by alternating courses of header (transverse) and stretcher (longitudinal) bricks (English bond, for strength) or by introducing half-cut bricks (Flemish bond, more attractive) (see Fig. 3.2).

3.1.2 Lime mortars

Mortar is the other important ingredient in the strength, durability and appearance of masonry, functioning as bedding agent to take out irregularities in the units and evening out load distribution and transmission from brick to brick. Technically, mortars are composed of a binder (traditionally lime or cement), aggregates (sand, crushed brick or fine gravel) and water. Once again, local availability of constituent materials determined the types, with lime being used as a binder where limestone could be collected and burnt. The availability of volcanic Pozzolana enabled the Romans to build stronger and faster and their joints could be made thicker (but not weaker) by adding brick dust.

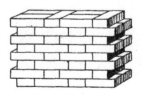 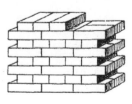

3.2
Examples of brickwork bonds: Flemish bond on the left, English bond on the right (Barberot 1895)

Lime is the preferable traditional binder as its mechanical properties are compatible to stone or brick and it provides the joint with the ability
to 'breathe' and flexibility that results from the fact that any small cracks
can 'heal' by re-carbonation. The lime cycle explains why (Fig. 3.3). Lime is
produced by burning limestone (rich in $CaCO_3$) in a kiln to remove CO_2 and
form quicklime (CaO). It is then mixed with water in an exothermic reaction
(generating heat) to produce lime putty or $Ca(OH)_2$ (slaking). During hardening, atmospheric CO_2 is absorbed (carbonation), a very slow process that
virtually transforms the putty once again into a form similar to stone.

Nowadays lime is available in two types. Non-hydraulic lime
comes as a putty (suitable for internal plastering) or dry powder (used as an
additive to cement mortars) and reacts with CO_2 in the air (carbonation) to
harden. Hydraulic limes result from limestone containing clay or silt which,
when they react with water, form calcium silicates and aluminates that set
and harden. This hydraulicity allows lime to handle moisture.

In areas where limestone would not produce sufficiently hydraulic natural limes, additives were sought that would cause setting and improve
strength, like Pozzolans, brick dust or furnace ashes. It is very important
however to underline that carbonation and eventually the achievement of the
desired strength is a very slow process. Careful construction planning was
often required, especially for those structures that depended on a formwork
(like vaults) and it is often a factor more important than ultimate strength
in determining a form. Cement (which is also lime-based) eventually managed to accelerate this process but eventually killed the production of lime
in some countries (UK). Natural lime is however a modern building material
and standards and designations exist to specify mortar often according to
hydraulicity.

3.1.3 Stonework

Stone is a natural building material and has three main types of origin: igneous (granite, basalt, gneiss), sedimentary (sandstone, limestone, poros) and
metamorphic (tuff, slate, marble). The geological presence of each type
would greatly influence the architecture and tectonics of an area even in

3.3
The lime cycle

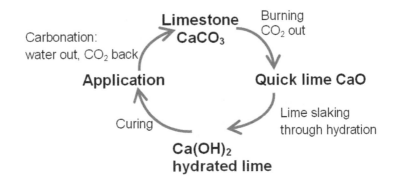

distinct periods. The classical Greek architecture of Olympia, for example, is characterised by local poros that is in contrast to the finer aesthetic, almost sculptural, qualities of the Pentelic marble used in the Athenian Acropolis and actually established the image of this style. The Romans used poorer tuff for their monuments until they captured the quarries that produced the Carrara marble. Most Gothic cathedrals in France and Spain are in limestone while sandstone was largely used all over Britain.

Rock of volcanic origin is formed by molten magma cooling and hardening at various rates. The faster the rate the smaller the crystals become, the finer the grain and the lower the strength. Granite formed by coarse grain is a very durable material and characterises the areas where it is available, such as Aberdeen, but is also harder to work with compared to sandstone. Its strength combined with the fact that it can be given a polished finish makes it an ideal cladding or even expensive building material. Hadrian, for example, used monolithic Egyptian granite columns at the portico he added during his restoration of the Pantheon in Rome, mostly as a symbol of his power to employ materials from anywhere in the world. Gneiss is a metamorphic form of igneous rocks, and types like the pink Lewisian Gneiss, one of the oldest rocks in the world, have been sourced locally for the construction of Iron Age brochs in the Sutherland area of Scotland.

Sedimentary rocks such as sandstone are formed by compression of muds and silts where the quartz in the sand is bonded by the contained clay or silica. Layers are formed and their density or changes in colour due to inclusion of other minerals give each type of sandstone a characteristic texture. Red sandstones (containing iron oxides) have been used in Scotland for some major public buildings (Portrait Gallery in Edinburgh, Kelvingrove Museum in Glasgow) while many of the prestigious neoclassical houses in Edinburgh were made from the finely grained and uniformly pale Craigleith sandstone (quarried locally).

Limestones or dolomites are the product of sedimentation of calcium carbonate ($CaCO_3$), the same material as natural lime, originating from organic deposits (from shells or bones), often occurring at the bottom of ancient oceans. They are versatile stones, which balance well porosity, strength, stiffness or durability. They are encountered almost everywhere and are not too hard to work with.

Limestones have distinct architectural qualities that contribute to the specific character of a building or an area (e.g. Paris). Major buildings from as early as the pyramids were built from the naturally occurring limestone. Portland stone is a buff colour type found in the south of England and is the material of St Paul's Cathedral, the British Museum and most of the emblematic buildings in London (such as the offices and residences along Regent Street). Chalk limestones have been used in the northern French cathedrals like Beauvais or Amiens, giving them the distinctive white colour of their façades. Pale oolitic limestone was used in Lincoln Cathedral and is still quarried today from a quarry in the Cathedral's ownership.

Depending on the rate of sedimentation and precipitation of the all-important calcium carbonate, other more porous limestones have been created. Travertine occurs where hot springs or stalactites have formed and the process results in the characteristic pitted surface (the density of the holes determines its use) and pale cream textures. It is easy to cut but can also be polished, so it was used as a load-bearing material, pavement or cladding. The Romans used it extensively for the structure of many buildings (calling it *lapis tiburtinus*), while it was also the material for many Baroque façades or modernist buildings later, especially prized for its ability to create a rather matt finish at a large scale.

Other sedimentary types can originate from deposits of other materials. Tuff was created from the consolidation of volcanic ash. It is a relatively soft and often quite porous stone that is easy to carve, but because of its high porosity requires direct or indirect protection from the elements, especially water. Many Roman buildings have a tuff fabric (*lapis albanus*) plastered or clad with more precious materials and many exposed load-bearing walls were built throughout history in Italy using types like yellow tuff, peperino, Riano, etc. in colours including grey, buff, red or green. Poros stone is a soft, easily worked calcareous sandstone or the result from sedimentation of shells, but the name often covers very porous limestones originating from marine or alluvial deposits. It gives the characteristic qualities in places like ancient Olympia, Corinth and Aegina but often was used only for foundations.

Marble is a metamorphic rock formed by the recrystallisation of carbonate minerals originating from sedimentary limestones. It has always been considered as a high-quality material that can be polished, because of its fine and homogeneous colours and textures, as well as its durability during carving and service life. Marble has been used mostly in sculpture and, by association and due to its strength, in representative public buildings. Famous white types include Parian or Pentelic marbles in Greece (used for the structure of the Athenian Acropolis) and Carrara (*marmor lunensis*) used in the Pantheon or the Caius Cestius Pyramid in Rome. Coloured types have also been used mostly for pavements or cladding, like pavonazzetto (a breccia-type rock formed by fragments of stone such as marble or limestones within a natural cement of a contrasting colour, creating colour variety), rosso or verde antico (reused), cipollino (green and white streaks), serpentine, and occasionally as columns in small prestigious monuments, such as altars or tombs.

3.1.4 Structural aspects in masonry

Masonry, depending on the bond or layout (Fig. 3.4), can have pronounced orthotropic properties in strength, elasticity, thermal expansion, resistance to fire, etc. and its characterisation is affected by the function of the joints as planes of weakness. The compressive strength of masonry is related to the strength of the constituent units (stone, brick, earth) combined with the

3.4
Typical cross-section
of a wall with facing
units encasing a
rubble core (tomb at
Via Bisignano, at Via
Appia Nuova, Rome)

thickness and form of the joints. The following tables summarise mechanical characteristics of types of stone and bricks, as they are found in historic masonry types. The data come from various sources and databases (e.g. McMillan *et al.* 1999; Esposito 1998).

Table 3.1 Mechanical properties of stone and bricks

	Density (kN/m3)	E (GPa)	fc (MPa)	Porosity %	Absorption % by wt
Binny	21.6	16.4	24.4	16.2	11.6
Craigleith	22.2		94.3	13.5	6.8
Cullalo	21.6	20.6	35.7	18.4	11.2
Hawkhill	23.7	27.1	42.8	10.1	7.1
Ravelston	26.3	40.4	64.4	3.6	2.6
Coal measure	23.0	25.0			
Lecce stone	26.96	1.24	10.8		
Kenchreae poros	15.3	2.0	4.2		
Riano	13.75	5.2	9.76	46.1	23.3
Neapolitan yellow tuff	22.94	0.96	4.86		
Peperino	19.3	2.81	5.78	12.3	2.6
Portland stone	22.0		46.0	17.6	5.5
Caliza de Hontoria	21.0	7.5	22.57	4.0	
Oolitic limestone	20.0		8.16	25.6	9.2
Marble	25.9	3.5–10.7	52	0.8	0.2
Travertine	24.8	55.6	90.7	5.0–12.0	0.89
Flagstone (Caithness)	27.0		159.0	0.3	
Granite	27.5	50.0	135.0	2.9	0.31
Gneiss	25.0–28.0	30.0–80.0	175.0	1.16	0.37
Basalt	27.5–31.0	100.0	330.0	7.0	
Modern fired clay bricks	22.0–28.0		50.0–70.0	7.0–36.0	4.5–7.0
Mud bricks	17.0–22.0		2.0–3.0		12.0

It is important, however, to remember that, as with modern masonry, such values cannot be considered to represent the strength or performance of masonry. Tests have been done separately, either in situ (using jacks) or in laboratory (on wallettes) to estimate approximately the strength of various masonry types in axial (compressive) loads, bending or shear. As Table 3.2 shows, there is a wide scatter and even less direct relation to the proportions of the constituents. Because of variety in typologies, conservation of materials, workmanship quality and many other factors, knowledge on the exact effect of the bond is inconclusive so a wider context of examples of orthotropic properties of masonry types from case studies or scientific research is presented.

For modern brickwork the relationship between the shape of the units, their strength, that of the mortar and the strength of brickwork is well established and implemented into design recommendations (Hendry *et al.* 1997; Karantoni 2004). Table 3.3 and Figure 3.5 show clearly that weaker mortars (due to low strength or high thickness, as is found in some medieval or vernacular types) significantly reduce the strength of masonry. Also, the strength of masonry increases with unit height (for a given strength of masonry unit). Such facts can be used as qualitative observations in the

Table 3.2 Mechanical properties of historic stonework and brick masonry (directions are along the bed joint 'Long.' and perpendicular or normal, 'Perp.')

Type of masonry	Unit weight (kN/m³)	f_c (MPa) Long.	Perp.	E (Gpa) Long.	Perp.	G (GPa)
Campanian tuff. T1: part filled cavity wall			0.99–1.2		0.56–1.28	0.25–0.66
T2: solid tuff wall			1.23–1.53		0.99–1.11	0.46–0.52
T3: wall T1 grouted			1.47–1.69		0.97–1.46	0.46–0.68
Pavia masonry panels (Berto *et al.* 2002)	7.61		3.0	2.99	2.0	0.9
Fossanova church (De Matteis and Mazzolani 2010)			5.0		4.2	
Wallettes, Dafni (Miltiadou-Fezans 2008)	21.0		1.74–2.26		1.0–1.5	
Tuff masonry (Augenti and Parisi 2009)			1.90		2.09	0.36 (code)
Historic brickwork (Zimmermann *et al.* 2010)			6.1		1.795	
Parliament Hill (Sorour *et al.* 2009)					0.35	0.18
Modern brick wallettes (Sinha *et al.* 1997)				15.5	10.69	
Modern walls (Ganz and Thürlimann 1982)		1.87	7.61	5.46	2.46	1.13

Table 3.3 Compressive strength of modern brickwork in relation to mortar composition (after Hendry *et al.* 1997)

Mortar description	Proportion by volume Cement:lime:sand	Mean compressive strength (N/mm2) Lab tests	Site tests
i	1:0–0.25:3	16.0	11.0
ii	1:0.5:4–4.5	6.5	4.5
iii	1:1:5–6	3.6	2.5
iv	1:2:8–9	1.5	1.0

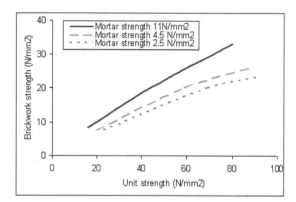

3.5
Compressive strength of masonry (after Hendry *et al.* 1997)

assessment of the strength and effect of the constituents comparatively, for example, within the same building among different batches of fabric.

Figure 3.5 also shows that the overall compressive strength of masonry is lower than that of its constituent units, by an average of one third. A similar observation can be made for other mechanical properties (elastic modulus, shear), while it has to be remembered that masonry is an orthotropic material (and irregular historic types are even anisotropic). There is a constant problem in technical literature, when shortcuts are taken, to use the properties of brick or stone to represent that of the masonry, usually due to the difficulty of performing tests to wallettes or specimens extracted from the fabric using invasive techniques. Once the moduli of elasticity are known in each of the main directions of the material, the shear modulus G_{12} can be evaluated as:

$$\frac{1}{G_{12}} = \frac{1}{E_1} + \frac{1}{E_2} + \frac{2 \cdot v_{12}}{E_2}$$

where E_1 is the elastic modulus in the stronger direction (along the bed joint) and E_2 is in the normal direction (perpendicular to the joints).

There is a great variety of configurations in the bond types or joints and often shear strength depends on the confinement conditions and the values of compression forces. For structural analysis, however, many of these properties might not be of use. The randomness in rubble types, the variety of stone strength and, above all, the presence of cracks and voids result in a wide variety of elasticity and strength of the masonry. This fact, together with the brittle character of the constituent materials make the use of elastic or solid continuum models often obsolete, and limit-state analysis becomes relevant as it gives information on residual strength, often the most essential property (see Section 2.5). For more intact structures these properties might be less spurious and therefore more useful, in which case continuum analysis could give some reliable information on the distribution of loads and the areas of maximum deformation.

3.1.5 The pathology of brick and stonework

The discussion of the pathology and durability of masonry will be focused on those aspects that affect a project of intervention in the stage of initial design. This covers the assessment of strength but also the tectonics and materiality of the original fabric, which eventually defines its character. Regarding mechanical failure, traditionally high strength reserves or redundancies in uniaxial stress (compression) and in-plane forces (shear) are expected because of conservative construction. Similar to modern practice, historic masonry structures can be under a combination of these forces with bending, but even 3D aspects can exist due to the thickness of the fabric and irregular variation (Karantoni 2004).

The stability of masonry under vertical loads should address not only the compressive strength but also the buckling of the wall. The latter is often an overall design issue and, as will be seen in more detail in Chapter 5, it may be the result of slenderness or eccentrically applied loads. Strength is directly affected by the constituents, geometry of units and pattern, and in this scheme mortar has the role of a soft plane confined between the stiffer units, or a plane of weakness in mechanical terms. When vertical load is applied the joint will expand more than the brick causing a vertical split in the brick rather than crushing.

The strength of brickwork in biaxial bending has been evaluated for various combinations of applied forces and becomes even more complex according to the stonework type. Often some of the criteria result from a phenomenological process, based on tests on specific typologies. The purely theoretical approaches range usually between the computationally demanding micro-modelling, and macro-modelling which suits better the macro-scale analysis usually occurring in the design of the interventions discussed in this book. Homogenisation techniques attempt to incorporate into a single constitutive law the properties of the constituents and the geometry of the masonry. A further complexity is certainly due to the need to validate the theory with tests and ensure the wide application.

The example of a plain biaxial moment criterion for modern brickwork (Sinha *et al.* 1997) gives insight into the orthotropic homogenisation problem and the failure patterns (Fig. 3.6). Cracking occurs when the applied moment reaches the moment of resistance in that direction but the key points are how failure spreads and is shared afterwards. For stress regimes

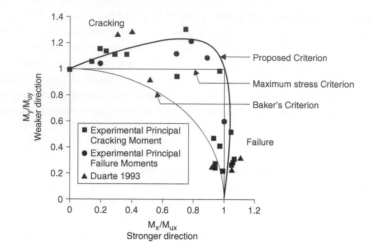

3.6
Biaxial failure
criterion on brickwork
(Sinha *et al.* 1997)

lying to the left of the point (1, 1), the weaker direction (normal to bed joint) will crack first, shedding the load to the stronger one. If ultimate strength in that direction is subsequently exceeded, failure will occur. When, however, the stronger direction cracks first (points on the right of (1, 0), the weaker direction cannot support the amount of the load transferred, so failure happens immediately. The load transferred to the neighbouring points can propagate failure until a sufficient number of fracture lines has transformed the structure into a mechanism (ultimate limit state), which collapses.

Concerning shear failure, as with every plate structure under in-plane loads the characteristic shear strength of masonry f_{vk} depends not only upon the (initial) shear strength of the masonry f_{vk0} (as evaluated from tests) but also the vertical applied stress σ.

$$f_{vk} = 0.4\sigma + f_{vk0}$$

Shear failure can be more evident in rather long walls where the bending effect from the in-plane force is reduced and cracks will appear along the diagonal facing the force (Fig. 3.7). Differential settlement is often a reason for shear failure, while seismic action would rock the plane, causing X-like damage within the plane or in areas of high shear concentration like the corners of windows or thickness change.

Such mechanical problems and their combinations result from high vertical or horizontal loads, as well as conditions such as imposed deformations (differential settlement or confined thermal expansion), excessive lateral forces (high wind loads or unconfined diaphragm action from adjacent floors) or even eccentric or offset loading due to failures in other parts.

Fire-proofing occurs naturally in masonry construction, as stone resists fire and bricks are produced in kilns. It is usually the deterioration of ancillary elements such as ties, reinforcement or pins holding ashlar blocks or cladding panels that are affected (see the failures of marble cladding in

3.7
Superimposition of
compression and
shear on a small wall
(Fouchal *et al.* 2009)

Cells
Sxy

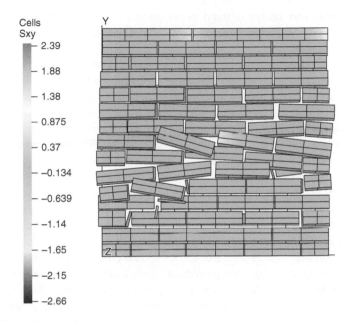

- 2.39
- 1.88
- 1.38
- 0.875
- 0.37
- −0.134
- −0.639
- −1.14
- −1.65
- −2.15
- −2.66

Roman public buildings). The overall stability is compromised by the failure of other areas in the building, such as the floors or roof and the fire engineering strategy must aim in reducing the possibility of fire reaching those areas, as discussed later.

The pathology of brickwork and stonework can be separated into factors pertinent to the original quality of the material, the presence of water, biological or atmospheric agents, and actions related to service (changes in use or loads, and accidental loads). The severity of the problems often depend on the porosity of the stone, individually and in relation to that of mortar, as it eventually controls the stone's capillarity and water permeability (see, for example, Dimes and Ashurst (1998) and Henry (2006).

Historic masonry often has more pressing durability problems, due to long-term exposure to the elements or low and variable quality of the fabric, both in elevation and thickness. Masonry is hygroscopic as the units can absorb between 4 and 35 per cent of their dry weight in water. When water infiltrates the pores, soluble salts or adverse atmospheric and biological agents can be deposited. The moisture content is therefore very important as it conditions the principles of deterioration and the transportation of adverse agents. Clays can also be a source of problems, either because of soluble salts they may contain when used for brick fabrication, or volcanic clay inclusions in sandstone that deteriorate at a different rate.

A similar mechanical deterioration of the pores occurs under sulphate attack. In bricks, it results from a reaction between the soluble sulphates they contain and tricalcium aluminate (present in cement when used in mortar repairs), producing calcium sulphoaluminate which expands. The reaction occurs under significant water movement that dissolves these

49

sulphates and transports them to the cement. Together with mechanical incompatibility, these are reasons why cement has to be avoided in historic masonry repairs.

In a freeze/thaw cycle, water or salt crystals can expand and contract causing micro-fissures. Rainwater penetration often affects a superficial layer and the problems manifest as spalling. Salts and sulphates migrate and deposit in this layer or the mortar joints (especially in thick joints, made of low-quality set mortar) and can expand by up to 9 per cent of their volume, harden, or further expand thermally. A gradual delaminating occurs which accelerates once inner and less seasoned layers are exposed.

A milder version of the problem is efflorescence, a white (occasionally green) soft deposit on the face of bricks. When water moves out of masonry it can transport salts that are naturally bound to the stone or present in the cement used in mortar. It is, however, superficial and, therefore, an aesthetic problem. More problematic staining in stone can occur when atmospheric conditions or surface texture provide sufficient water for biological growth. Algae can grow quite rapidly creating a typical green stain that allows further contaminants to accumulate and create a tough patina. Lichens, on the other hand, can grow more substantially (up to 35 times their volume in a very wet climate) and discharge acids on the masonry surface. The extent of staining depends on to what extent rainwater or wind can remove the contaminants. Stains can also occur when rust from decaying iron or steel elements is washed onto a surface, deposited at a rate depending on the roughness of the texture.

More serious stone decay can result, however, from atmospheric or biological pollution. Atmospheric sulphur dioxide (SO_2) produced from the burning of fossil fuels (petrol) or deposited through acid rain, converts to sulphuric acid (H_2SO_4) under rainfall, which in turn dissolves the calcium carbonate of limestones, converting it to calcium sulphate, i.e. crumbling gypsum.

Extensive loss of fabric can occur due to decay. In combination with the inner structure of the stone and the stress regime, fissures that were created due to crystallisation can extend and cause local failures. Such a process was considered to have been accelerated by traffic vibration in the case of the Civic Tower in Pavia, resulting in a sudden collapse that had given no warning signs. Another important problem results from incompatible repairs, often when cement is used. Hard cementitious mortar used in re-pointing can induce decay of the adjacent units (whether brick or stone) as the mortar will then be more impermeable than the stone. The water movement, and associated decay, will happen within the (weaker or softer now) stone rather than through the joints, resulting in curious patterns where all the units could disappear leaving the joints intact.

Problems in modern construction related to bad design or execution are even stronger causes of deterioration in historic masonry. 'Inappropriate specification' or 'low understanding of physical behaviour of masonry and its constituents' are essentially the issues when successful traditional practices are ignored or improperly transmitted. This often happens as a result of social or political breakdown, as in the case of Roman structures being poorly imitated in Rome or elsewhere in Europe between the fourth and ninth centuries after the fall of the Western Roman Empire. 'Inadequate control during construction' is often found in remote sites built rapidly (castles) or long-lasting construction projects (like cathedrals) where a variety of master-masons and builders would follow their own standards if they were not skilled enough to learn from good practice on the same site.

3.1.6 Reinforced concrete

The discussion will deal with modern concrete, in its form as reinforced concrete (RC), which has been a mainstream building system since the 1870s and therefore covers many historic buildings, especially those of the modernist movement. Concrete and steel are two systems where technical development especially in the twentieth century has clearly influenced their effectiveness, so it is important to make close reference to a timeline and key buildings when dealing with properties and pathology.

The two key properties that initially attracted designers were the prefabrication of elements and the ability to imitate stone, especially after Portland Cement was patented in 1824 by Joseph Aspdin, following success as a material for garden sheds. Volume and higher compressive strength was achieved when the Roman form of 'opus caementitium' was replicated by introducing sizeable aggregates and RC was eventually patented in 1867 by Joseph Monier to successfully address the low tensile strength of the conglomerate.

Compared to stone, the stress–strain properties show a relatively higher strength, and a plastic region after yield strength is exceeded, which allows some ductility and redistribution of loads, in contrast to the brittle nature of stone (Fig. 3.8). The ductility and tensile strength are further improved when steel reinforcement is introduced as bars (in bending) or stirrups (in shear).

3.8
Stress–strain curve
for concrete and steel

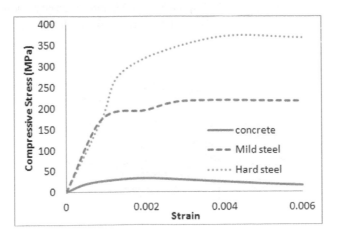

Other important features, in comparison to more traditional or natural materials, are that concrete can be cast into a variety of shapes and most of its strength develops seven days after casting. This is produced by mixing cement with water, causing its compounds (but not the aggregates) to react chemically. Concrete then hydrates until it starts to harden and set through a carbonation process similar in principle to that of lime. The chemical reactions are accompanied by release of heat of hydration but full setting is a process that can take even years.

Most of the early applications of concrete were in prefabricated building elements, as in the interesting example of Castle (or Portland) House built in 1851 as a showcase of concrete products by John Board in England. Some early concrete structures are actually boats, a technique perfected later by P. L. Nervi in his ferrocement yachts.

The potential for frame buildings and industrialisation was understood from the 1880s following a slow move from precast elements to monolithic construction. Inevitably the initial influence was iron and timber, imitating their construction processes, but soon proprietary systems were developed: Beton Coignet (France, 1870), Ransome Stone (US, 1884), Kahn, Demay-Freres (1887), Hennebique (Belgium, 1892), von Thullie (1897). The Hennebique system (Fig. 3.9), with its influence from steel, somehow established the modern practice, either because of the effectiveness of its reinforcement details (use of metal plates, uneven reinforcement for shear), or due to intensive (almost ruthless) marketing and dissemination by its inventor. Before the establishment of codes of practice in the 1920s, designers would buy the system of their choice from a local licensee, who would size the structural elements and specify the reinforcement.

The RC frame spread from Germany (*Stahlbeton*) and France (*beton armé*) with the first significant buildings in the 1890s, including: apartments in rue Danton 1, Paris (1892), Weaver's Mill in Swansea, UK (1897), Roeder Printworks in Leipzig (1897), Genoa Port Silos (1901), Ingalls Building, Cincinnati (the first skyscraper built from RC, 1903), office block in Syntagma Square in Athens (1909). Theory and practice were taught in the École des Ponts et Chaussées in Paris as early as 1897, and specialist publications have been in existence since 1898 (Hennebique's *Beton Armé*).

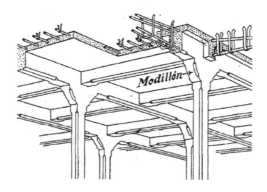

3.9
Details of the
Hennebique system
(Barberot 1895)

Further key developments were either new techniques (such as Freyssinet's invention of pre-stress following the repair of Le Veurdre concrete bridge in Vichy in 1907) or the presence of influential designers such as Robert Maillart (1872–1940) or Fritz Leonhardt (1909–99) on bridges, Auguste Perret (1874–1954) on buildings, Eduardo Torroja (1889–1961) on shells, P. L. Nervi (1891–1979) on standardisation, or Owen Williams (1890–1969) on infrastructure. Each had different professional or academic backgrounds and their own attitudes towards construction and concerns about expression of form.

Modernist architecture placed concrete in an increasingly central role, associating its versatility and physical properties with many of its expressions, such as functionalism or brutalism. Robust skeletal forms or exposed *beton brut* supported the expression of simplicity or strict relation to the planned function of the building on different scales and by various designers (Louis Kahn, Peter Womersley, Le Corbusier, G. Terragni, F. L. Wright, M. Fissac, O. Niemeyer, etc.). Later developments, such as prefabrication, allowed a more industrialised production of concrete buildings, especially multi-storey housing projects or infrastructure.

There are three key aspects that control the application of concrete and where most interventions focus: strength, surface finish and construction process. RC structures are relatively new so they were designed following a scientific process and safety approach with a similar rigour to modern practice. Structural problems would then result from degradation of the material, incorrect original detailing, design errors or intermediate alterations. Certainly these are common problems even to many modern structures that have been 'designed', i.e. their layouts and member sizes have been determined using an analytical process and verified through building codes.

The effect of material degradation is primarily manifested in the inability of concrete to protect the reinforcement from corrosion or fire and protection usually focuses in the cover (Fig. 3.10). Decay is mainly associated with carbonation or the loss of passivating alkalinity in concrete that protects the steel bars against acids: atmospheric CO_2 dissolves in water (in concrete's early liquid form or in the pores) to form a mildly acidic solution.

3.10
Visual classification of defects and signs of pathology in RC structures

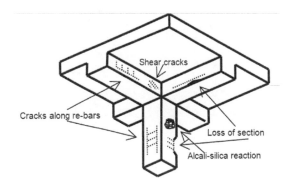

If CO_2 increases, then the pH drops (acidity increases) and damages the passivation function. The carbonation (Fig. 3.11) starts from the surface layers (like in stone), moving as a front until it reaches steel. Reinforcement bars can then be attacked by moisture and oxygen, resulting in corrosion.

Exposure to aggressive chemical or chloride attack has a similar effect in causing the loss of passivating alkalinity. Its sources are basically salt, which comes from sea water (in marine environments), poorly washed sea-dredged aggregates or de-icing salts used in roads during snowfalls, a typical problem that affects all concrete decks in bridges built from the 1960s. A more inherent source is cast-in chlorides (calcium chloride) that were used in 1970s as accelerators.

Problems resulting from detailing, such as over-reinforcement, affect especially the joints. If spacing is smaller than the larger size of the aggregates, voids can be left, producing mechanically deficient sections and increased porosity. Quality and compaction of concrete, or cement content are also known to control porosity and the effectiveness of the cover. Low durability was a problem when RC became the norm in residential buildings: some post-war-designed mixes had lower cement contents and changes in the grinding of cement lowered the quality, resulting in an overall visual dissatisfaction that would worsen due to lack of maintenance.

Other material problems resulted from the declining strength of high alumina cement concrete (HAC or CAC), a mix designed for quick curing and very high durability and strength. HAC produces high early strength that reduces significantly afterwards because of 'conversion'. It was used widely for standard pretensioned floor joists until 1976 when it was banned after a failure. The reaction that followed is nowadays considered excessive and is indicative of the often conservative approach of the industry to risk.

Another better-known condition is Alkali-Silica Reaction (ASR), a chemical process: alkalis from cement combine with certain types of silica

3.11
Carbonation in RC

in the aggregate when moisture is present. An alkali-silica gel is produced, which absorbs water and expands to cause a map-cracking pattern which follows the lines of main reinforcement. This is a greater risk in bridges, hydraulic structures, exposed frames or foundations. The problem is not immediately structural but allows water to infiltrate, so it requires critical examination of the robustness of reinforcement and, eventually, limited strengthening or partial/full demolition followed by rebuilding.

Early concrete was also reinforced with smooth bars, which offer less grip to the concrete, while the theory was not developed enough to provide resistance to the full range of loads. Moreover, the original design codes may have made insufficient provisions (or none at all) for certain loadings or phenomena (e.g. accidental damage, increased imposed loads).

There are also conditions that affect all modern construction systems with a long service life: apart from changes in loads, the soil conditions may have been affected by changes in the water table, subsidence, contamination, etc., or by significant alterations in load distribution, causing differential shear failures. Creep and loss of strength due to carbonation can increase deformations and affect serviceability, a problem that could even cause the loads of slabs to be transferred onto the window frames in openings (as occurred in the cases of Fallingwater and Boots D10, which will be examined in more detail later).

Disproportionate collapse is a problem that largely affects prefabricated buildings. This is a structural issue essentially caused by the lack of design for robustness, where insufficient vertical ties were provided, relying on the effect of the weight of the large panels for the stability of the building (often the case in the UK), almost as an extrapolation of historic masonry technology. The weakness becomes apparent when the ties deteriorate due to water ingress, often caused by bad jointing, and corrosion occurs in the reinforcement, which is also exacerbated by spalling of the panels resulting from the low durability of concrete and inadequate specifications.

A widely applied system was the Larsen Neilsen (Denmark 1948), as adapted in UK by Taylor Woodrow – Anglian Ltd (TWA), with the first blocks being built in Newham, London in 1965. Other systems include Laings' 12M Jespersen (Aylesbury Estate, Southwark), Wates or Bison (Glendinning and Muthesius 1994). The problem of lack of vertical tying became prominent at the Ronan Point collapse, which damaged a 22-storey tower block in London, on 16 May 1968, after a gas explosion. Progressive collapse of an entire corner of the block followed the destruction of a single flat as the connection between the large panel precast concrete floors and walls relied on friction only, with no ties. The outcry (four casualties) caused radical changes in regulations ensuring redundancies were incorporated.

Finally, as far as fire resistance is concerned, concrete itself is non-combustible, but its strength deteriorates under prolonged intensive heat. Steel (Fig. 3.12) loses strength rapidly above 500 °C (a room fire is between 800 and 1000 °C) so the key problem is the resulting redistribution of loads following plastic failure of the sections.

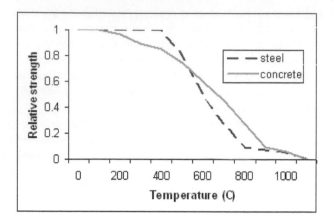

3.12
Reduction of strength
in RC and steel with
fire

3.1.7 Steel and iron

The evolution from iron to steel is about enhancing properties such as high strength (in both compression and tension) or stiffness through the control of carbon content and admixtures, while guaranteeing industrialised production (quality and quantities). Carbon guarantees compressive strength but makes iron brittle, while admixtures can improve performance, including durability. As in concrete, there is a chronological development marked by technological breakthroughs in production and key buildings (Fig. 3.13).

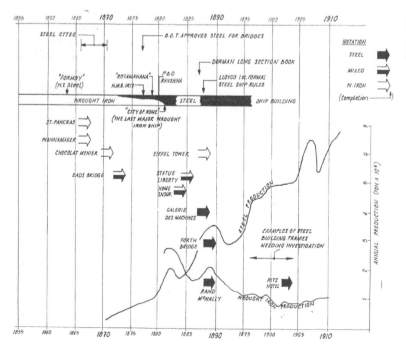

3.13
Evolution from iron
to steel structures
(Sutherland 2000)

Iron has always been used in small quantities for specific applications, for example, the original high-quality bars that held together the drums in the columns of the Acropolis monuments in Athens. Then considered as a precious material, it was cast iron that became available in larger and sustainable quantities in the eighteenth century and showed structural potential, initially in bridges. In essence, it is the product of extracting iron from ore by removing impurities (pig iron) and keeping about 1.8–5 per cent of carbon and small amounts of sulphur, phosphorus and silicon. One of its main forms is the grey type (graphite flakes), which is hard and brittle, but easily cast, with high compressive strength and resistance to corrosion. It is non-combustible but loses strength on heating so it is often clad and fire-proofed by earthenware pots in plaster.

Key applications of cast iron have been the 1779 Ironbridge (spanning 31 metres over the river Severn with a clear influence from timber joinery in the connections and masonry arch spandrel forms) or the 1795–1805 aqueduct at Pontcysyllte by Thomas Telford (1757–1834) and William Jessop (1745–1814), where a cast-iron trough is mounted on iron arched ribs 13.8 m span × 2.3 m rise. Cast-iron members of the period are distinguished in their rough texture (a result of casting). The relatively low strength in tension limited the spans beams could reach, or their bottom flange would be made wider or deeper to resist the higher moments at mid-span.

High tensile strength and resistance to fatigue were achieved by completely removing carbon and impurities, producing wrought iron. Also known as 'malleable cast iron', it was a refined process of rolling cast iron that removes all carbon, resulting in the purest form of iron (1783) and dominated structural frames between 1850–90. In nineteenth century, wrought iron was used as a flexural beam element, combined with cast-iron columns or arches (receiving the compression forces), appearing gradually in major public works and buildings: roof trusses at Euston Station, London (1837), ties in bridges (Newcastle 1849) or floor trusses (Crystal Palace 1851).

Controlling the carbon content, the effect of additives, and mass production were the essential next steps as construction volume increased. Essentially, steel is a developed form between cast and wrought iron, produced by the addition of other elements (Mn, Cr, Mo, Ni, Cu) and adequate heat-treatment, increasing strength significantly. The process became possible when furnaces were refined to control the carbon content, like Henry

3.14
A cast-iron column
connected to a
riveted beam
in wrought iron
(Barberot 1895)

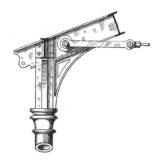

Bessemer's converter in 1856 or C. W. Siemens patent (1860s). Other key properties include good ductility, fracture toughness, weldability, high resistance in fire and corrosion.

Mixed structures were designed initially, like St Louis Bridge, Mississippi (James Eads 1874) or the steel wire suspension cables used in the 486 m-span Brooklyn Bridge, by John Augustus Roebling in 1883. The Home Insurance Building, Chicago by William Le Baron Jenney in 1884–85 was the first skyscraper in complete skeletal steel framework, starting the tradition of similar buildings in Chicago. Another new typology that was developed at the time was the department store (e.g. the Rinascente in Rome, by De Angelis & Bucciarelli 1885–87), while the decorative vocabulary that was established earlier by wrought iron found expression in Art Nouveau, for example, in Victor Horta's buildings in Brussels (Maison du Peuple, 1896–99; L'Innovation, 1901).

The image of steel, associated with transparency and fine lines, took time to establish as initially steel skeletons had to be clad in ceramic or concrete panels due to conservative fire regulations. The gradual disintegration of the stiff vocabulary is marked in the Reliance Building (Bernam & Co., Chicago 1890–94), where windows occupy the maximum space available by regulations and technology. The vocabulary of the curtain wall was explored by W. Gropius in Fagus factory, Alfeld as early as 1911, but it is only imagery as there are still large load-bearing walls. It was mainly L. Mies van der Rohe who consolidated transparency and lightness in his experimental Barcelona Pavilion (1929) or later in the USA at the Crown Hall, IIT Chicago (1956), which marked the development of corporate American vocabulary.

The pathology of the system has specific structural manifestations (such as in robustness and stability of the building or integrity of connections), which will be dealt with as part of the behaviour of frames in the next section. Other material problems include low structural capacity of material (defects, increased loads, corrosion, fire) or inadequate long-term durability. Sometimes it is simply an update to current safety levels (fire, increased capacity) that may require an intervention.

Corrosion results when steel reacts with oxygen and/or water, producing forms of iron oxide (rust). Essentially, a 'battery' is formed (Fig. 3.15) that transports electrons from iron in the sacrificed anode (which is corroded), building up material (rust) to the cathode.

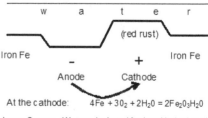

3.15
Corrosion mechanism in steel

The rate of corrosion depends on climate as well as local environmental factors such as exposure of steel and the humidity of the environment. Corrosion is also accelerated by certain salts, a problem that attacks many steel structures in a marine environment or those requiring de-icing salts, especially bridges. For example, indicative corrosion rates for steel and iron, expressed in μm per year, are:

dry rural climate	0.25
marine and industrial	80
tropical marine	620

A similar 'battery' effect occurs also when steel is in contact with other metals or alloys, such as nickel and copper, causing galvanic or bimetallic corrosion. The rate depends on the material and steel will corrode in contact with a more noble (resistant) metal (which becomes cathodic). The following ranking from noble to most corrosive (Galvanic Table) is based on their decreasing galvanic or electromotive force (EMF): graphite, gold, silver, titanium, stainless steel, nickel, copper, bronze, brasses, tin, chrome, lead, steel, aluminium alloys, zinc, magnesium. Steel columns, for example, should not be in direct contact with copper cladding and a gasket needs to be used in the connection, which should not be graphite.

The different types of iron and steel are also distinguished by their pathology. Cast iron is brittle and cracks permanently without any warnings, while the more malleable wrought iron delaminates.

It is very important to consider the key characteristics of the mechanical behaviour of steel when assessing the design of historic steelwork, i.e. high strength and ductility (Fig. 3.8): after yield failure, the plasticity of steel allows loads to still be carried through the section and the increased amount to be redistributed to adjacent sections. A steel frame will eventually fail if a sufficient number of such failure points (or hinges) is created to transform the building into an unstable mechanism.

Fire is the other major action that affects the stability of frames. Steel does not burn but at 550 °C it loses 30–40 per cent of its strength and stiffness (Fig. 3.12). Fire action more rapidly affects the entire building, where variation of the fire load on each member causes them to fail at different rates. Metallurgical changes may also occur at high temperatures: steel can become further 'annealed', damaging the 'warm-working' effect of rolling (which prevents stresses from being built-in) and eventually reducing its yield strength.

The thinner sections and lower fabrication quality of early iron and steel meant deformations would be more pronounced, this is why fireproof solutions such as brick cladding or jack-arches were used until knowledge of fire safety was more confident. A more passive and effective approach followed today is the overall prevention of fire and fumes from reaching the vulnerable parts by designing structural and spatial compartmentation.

3.1.8 Timber

Timber can be a remarkably strong and stable material, when adverse environmental conditions are controlled, and high qualities are available. It is probably the first material that was used for human dwellings as it is naturally occurring and requires little processing (sawing, drying, preservation). Relative to other traditional systems, it has had application in the most extensive variety of structures, with a very efficient performance (good resistance to atmospheric agents and heat retention), combined with strong aesthetics. Intuitively, joiners have become aware of its high strength-to-weight ratio and, despite some catastrophic failures, its high fire resistance and stability.

The key aspects in the use of timber are the procurement of members of sufficient and length and straightness, consistent strength, acceptable textures and durability. Being also the most vernacular of the naturally occurring materials, the development was strongly craft based and dependent on the tools available. Structural stability was empirically assessed but was more direct in framed structures and further improved together with the technology of joints.

Strong timbers, produced from well-seasoned oak would be used for large spanning roofs of prestigious public buildings (such as the roofs in cathedrals and probably ancient temples). Consistent and straight beams and columns however would be available only in short lengths, which prompted strategies to make the best use of them, for example, coffered ceilings (often richly decorated), hammer beam roofs (Westminster Hall, Stirling Castle) or closely spaced stud frames in houses. The gradual introduction of stronger and more durable iron tools enabled higher precision in the sawing and jointing of timbers and the strength of joints was further enhanced later by the industrial production of nails.

Joints developed from plain ropes lashing together green branches (like the withy frames) or poles (in the Celtic Crannogs) to increasingly sophisticated notches such as dovetails (Fig. 3.16), dowels or mortice and tenon joints which appeared in medieval times. Transfer of tensile forces was a common structural problem and, apart from these techniques, most of the earlier solutions tackled it by transferring the forces in compression. The development of efficient tension joints was essential and scarfing joints were a first attempt to transfer flexural tension. With the capability of producing larger quantities of iron, eventually cast-iron shoes or tension rods were introduced.

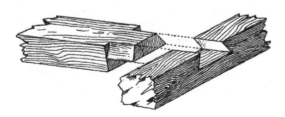

3.16
Dovetail joint
(Barberot 1895)

The development of structural types is better discussed in the next section. Constant improvement of forest policies since the Middle Ages meant timber production would be more sustainable, making more and healthier species available. As in stone masonry, waste is a major issue in procurement and sawmills have become more profitable by using most of the section of trunks. Better (and more scientific) understanding of properties and durability also enabled stress grading and optimised use of timber in specific areas of a building. Other techniques, such as reinforcement with metal lamination enabled larger spans and lengths (King's Cross Station, London 1852). Finally, a major contribution of timber structures is in the establishment of the technical and architectural vocabulary of early steel buildings (Ironbridge 1779, Bibliotheque Nationale, Paris 1854–75).

Strength often depends on the species, which today are divided largely into softwoods (more fibrous, of conifer origin, such as pine, Douglas fir, larch, spruce) and hardwoods (uniformly grained, from deciduous or evergreen trees such as oak and ash). The majority of hardwoods are stronger, stiffer and more durable than most softwoods and they have been more widely available historically. A wider discussion on the pathology and conservation of timber can be found in Ross (2002) and Yeomans (2003).

There are some inherent characteristics that weaken wood, including knots (function as stress concentration points), fissures or horizontal cracks along the grain, splits and shakes from drying, wane, resin pockets, etc., which nowadays are considered in the visual stress-grading process.

The durability of timber structures depends largely on the moisture content (MC). Many species will decay if kept at high MC for long periods and above 18 per cent MC timber is 'wet', attracting mould or staining. Rapid drying can cause shrinkage which, if differential, will produce distortions. When used internally, timber can be reached by moisture from leaking roofs and openings, or from the soil through inadequate damp proofing, and if the rooms are not continuously ventilated dampness will spread.

High MC in combination with certain ranges of temperature or light can also favour biological deterioration through growth of fungi, bacteria, lichens or algae. Fungal growth often manifests as dry rot (fungi growing above 30 per cent MC) or wet rot (50–60 per cent MC), requiring temperatures around 25 °C. Algae and lichens have similar mechanical and physical effects on wood as they do in stone (staining, rapid expansion, acid deposits). They can eventually degrade the cellular structure of timber, resulting in loss of strength, softening or disintegration and discoloration. Problems from insect infestation (woodworm, beetles) are mainly due to their larvae feeding on, and growing inside, the grain, affecting the structure of the timber.

There are also the usual service-related defects. Deformations can originate from creep or can be locked in from original drying of green timber. Instability results often from removal of essential timber members in the course of the life of the building or decay of the timber and connections.

Fire is a key problem but its effect is often unduly exaggerated. Timber is combustible but it is consumed slowly and superficially. It does not melt or distort and as geometric integrity is maintained sudden collapses occur only in very catastrophic and prolonged fires. Charred and uncharred sections are clearly defined so under a short fire a building can maintain some integrity of the original volume or essential elements like the beams and floors. Timber's fire resistance is independent of temperature (even at 1000 °C) and it retains its mechanical properties at high temperatures. Actually, the charred layers in a section insulate and protect the intact parts, which maintain their strength, therefore a timber structure can have longer periods of fire resistance (almost an hour in modern standards). It is often metal connectors or inadequate thermal insulation that allow fire to spread to the timber.

3.1.9 Soils and geotechnics

In contrast to the other aspects of historic building fabric, nothing much specific can be said about soil, except in the case of heavily deposited archaeological sites or later buildings on top of them. In general, foundations could be laid on a wide range of soils according not only to technical but also historic or cultural conditions, sometimes even in areas that would compromise their stability and not be allowed nowadays.

Rock is not ideal for foundations especially if strong tools are not available. Volcanic rock can be very hard: it does not allow for any foundations (e.g. around the basaltic Arthur's Seat in Edinburgh) or forms slippage

3.17
Stud frame building destroyed by fire in the Sultanahmet Quarter in Istanbul

planes for a structure (like the Clachtoll broch in Sutherland). At the other extreme, soft sandstone can not only hold a building, but occasionally allows cellars to be created (as in the elaborate hand-carved caves in Nottingham's Triassic sandstone). In some cases trenches could be cut to provide strip foundations and where the rock face was irregular a platform would complete the substructure (as in the Athenian Parthenon or the massive concrete platform of the Temple of Venus and Rome).

Typical soils composed of gravel, sands, silts and clay are expected to compress gradually and their compaction to fluctuate according to their permeability or the level of the water table. Clay consolidation should not be a crucial problem in long-standing structures, as soil should have settled earlier in the building's life. Changes in water content in combination with highly expansive clays can cause movement in the foundations. Increase of moisture produces swelling and upward movement of foundations (heave), while decrease causes shrinkage and increased settlement (subsidence). Up to 150 mm of subsidence is a typical problem, for example, with London Clay. Changes in immediate conditions (presence or removal of trees) also affect the water table and in combination with roots can cause differential settlement, resulting in tilting or shear failures.

The continuous tilting of the Tower of Pisa (stabilised since 1999 at 5°) is the product of ongoing consolidation of soft and variable estuarine deposits of silts containing sand and clay lying over a very sensitive layer of Pancone clay. Differential settlement has also been observed in cases where heavy buildings extend over variable soil conditions, such as the east apse of Lincoln Cathedral and the entire Metropolitan Cathedral of Mexico City, or subsidence in nineteenth-century tenements in Edinburgh and Amsterdam.

Other problems in soils can originate from changes in environmental conditions. Scour, i.o. the erosion of gravel and sand due to increase in river flows following high rainfall, often affects the supports of bridges, as happened in the first Pontypridd bridge or the recent collapse of Ponte de Fero over the Douro River (assisted by illegal sand extraction). Torrential rain can also have devastating effects, as happened in the collapse of the San Andres Martello tower outside Tenerife, when a torrent washed away the foundations under the north side of the tower. More recently (2009), the collapse of the City Archives in Cologne was the result of groundwater drawdown that disturbed the water pressures in the soil during pumping for a nearby metro tunnel, causing movement of the earth.

Archaeological soil has critical issues of stability, both during excavations and as a top soil layer in many modern cities. Earth and debris often keep ruins in a precarious position or even preserve them, as in the case of the volcanic ash and pumice that buried Akrotiri (in Santorini) and Pompeii. Warsaw was reconstructed after war damage on top of the rubble of the pre-war city (Lorenz 1964), in a similar way to the stratification of ancient Troy. Such sites would often contain unknown basements and voids that weaken new foundations.

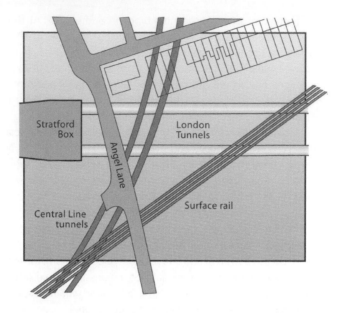

3.18
Overlapping of
underground lines
in London (Woods
2004 – courtesy *The
Arup Journal*)

Another typical problem of historic buildings in modern cities is the presence of underground structures, sometimes as extensive as tunnels and subways. The collapse of the Cologne Archives is a typical example of how delicate such interventions are, while in cities like London the density and variety of tunnels (existing and planned ones, from the Royal Mail to the CTRL tunnels) requires very careful study to avoid the overlapping of the zones of influence of soil pressures and their increase on underground structures (see Fig 3.18).

3.2 Building types

The problems outlined so far become even more complex in the context of buildings and the actions their design has to resist, so a review of types and the specific structural patterns and pathology is required to show the cross-links. So far, the effect of the pathology of materials has been discussed and there are several more sources of structural problems, such as original design issues, maintenance or even later restorations, conversions and changes in use. Regarding the structural issues themselves, problems can arise from vertical or lateral stability, failure of ties and joints, subsidence/heave or even insufficient fire-proofing.

3.2.1 Cellular (masonry) structures and the load-bearing wall

This is the typology most commonly associated with historic fabric. Structurally, these types are characterised by high in-plane strength, both in the vertical and horizontal sense (shear and compression). For stability, the walls must be assembled into cellular configuration, where the intersecting walls restrain each other. The scheme depends on the bond between the load-bearing walls, completed by the floors and roof functioning as stiff diaphragms. Some significant historic typologies will be studied in more detail as they represent examples of typical problems and possible interventions.

Structures in Roman times could be substantial and even reach four or five storeys, in private (*insulae* housing in Ostia and Trajan's Forum) or public buildings (Domus Tiberiana on the Palatine Hill, Tabularium). The high quality of Roman concrete was instrumental (Giuliani 1992) and thickness could be reduced to as little as one-eleventh of the height of a wall (as in the Pecile in Hadrian's Villa in Tivoli which measures 0.75 m thick × 8.3 m high × 214 m long). Establishing the structural vocabulary of almost all masonry construction, walls would be load-bearing or simple partitions, differing apparently in the quality of mortar or aggregates rather than the thickness. The load-bearing walls followed a strict cellular layout but their thickness reduced with height. In more prestigious buildings, a core in rubble or tuff was clad in marble plates or travertine blocks on one or both sides, with tying being the main concern. There would be rules for the location of the openings, which would be further strengthened with relieving arches.

Structural problems would arise during their lifetime due to poor quality of the materials or foundations, insufficient bond of the walls, or overloading. Fire affected the roofs and floors mainly, as the walls were incombustible, but a catastrophic fire would cause the collapse of the closely spaced joists and the gradual disintegration of the wall. Metal pins holding the cladding panels or functioning as ties would also melt. This would be more critical in walls made of larger blocks, such as piers of amphitheatres or city walls, which were fixed with pins and would often be vandalised for the precious metal of the joints.

Brickwork buildings would follow the same scheme throughout history with various stages of refinements in the structure of the wall or the floors. The design followed progressively a set of rules of thumb, like these examples from England that give the thickness of the walls:

- 1189: party walls 3 feet thick for 16 feet high, random rubble.
- 1611: for room height 10 feet, 2-storey – 15 inches thick; 3-storey – 18 inches thick.
- Piers not less than ½ width of windows.

More prescriptive recommendations appeared later from the London Building Acts of 1844 (CIRIA 1994). The building in Figure 3.19 for example, would be expected to have external walls 13.5 inches thick (0.34 m) at the lowest two floors while the upper one would be 9 inches thick (0.23 m). In later industrialised times, up to 10-storey-high buildings could be built in Chicago in 1870. The Victorian textile mills in England have load-bearing external walls and an internal skeleton of cast-iron columns. Floors use wrought-iron joists with a timber deck and the external wall provides lateral stability.

3.19
A typical nineteenth-
century brickwork
building in
Nottingham

These buildings often provide more combustible fuel for a fire in the floors and lightweight partitions made by lath and plaster. Lack of compartmentation allows fires to spread easily and as they grow well contained within the brick walls they can reach a point called 'flashover' where all remaining sources of fuel will ignite. The extent of destruction of the combustible elements can eventually make any temporary shoring or even reconstruction untenable, as was the case of the Cowgate fire in Edinburgh in 2002 that forced a very large block to be demolished. Dramatic historic fires have often caused demand for major changes in design and planning, for example, the great fires in Rome (64 AD), London (1666), Istanbul (Pera district 1870), Chicago 1871, which reduced density (downsizing projecting adjacent floors), promoted less timber construction and more fireproof materials, and enforced access for fire brigades etc.

Neoclassical stonework as developed from the eighteenth century detaches the fabric from its envelope. The load-bearing walls are made of coursed rubble and clad by discrete, well dressed stone blocks in ashlar pattern. The blocks are fixed with pins on the rubble core and they carry part of their load on the lower courses.

The choice of stone (such as sandstone in Edinburgh) is conditioned more by durability and ability to create fine lines and decoration rather than strength. The diagram by Bob Heath shown in Figure 3.20 (commissioned by the then New Town Conservation Committee, now Edinburgh

3.20
Typical cross-section
of a neoclassical stone
wall in Edinburgh and
indication of correct
and wrong use of
repair sandstone
(courtesy Bob Heath
and Edinburgh World
Heritage Trust (EWHT),
from Davey *et al.* 1981)

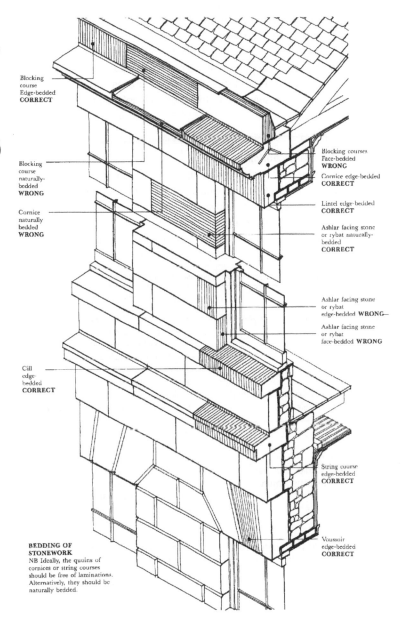

Blocking
course
Edge-bedded
CORRECT

Blocking
course
naturally-
bedded
WRONG

Cornice
naturally
bedded
WRONG

Cill
edge-
bedded
CORRECT

Blocking courses
Face-bedded
WRONG

Cornice edge-bedded
CORRECT

Lintel edge-bedded
CORRECT

Ashlar facing stone
or rybat naturally-
bedded
CORRECT

Ashlar facing stone
or rybat
edge-bedded **WRONG**

Ashlar facing stone
or rybat
face-bedded **WRONG**

String course
edge-bedded
CORRECT

Voussoir
edge-bedded
CORRECT

**BEDDING OF
STONEWORK**
NB Ideally, the quoins of
cornices or string courses
should be free of laminations.
Alternatively, they should be
naturally bedded.

World Heritage Trust) for example, has played a quite important role among conservation practitioners in Edinburgh as it shows how sedimentary blocks should be placed to avoid spalling decay. The construction process also becomes crucial, as both the ashlar leaf and rubble core have to be built in concurrent lifts and tie together with through blocks (inbands). The most load-bearing elements are lintels over openings and bay windows, but if care is taken in laying the blocks above openings an arch effect is created.

Problems that often occur are related with the failure of the ties or the mortar that fixes the stones. These are mostly serviceability failures that do not affect overall stability but falls from chimneys or coping stones from gables can cause fatal accidents as has happened in Edinburgh. A benefit of these disasters is, at least, that public, professionals and building owners became more sensitive to the need for the maintenance of these elements.

3.2.2 Timber frame buildings

Timber, RC and steel frame structures have to resist bending loads and should provide sufficient connections or stiffness to alter the distribution of moments favourably. Layouts and spans become crucial as they ultimately determine the size of beams and columns or the use of materials and strategies to resist tension. Timber frame, specifically, offers a wide range of

Table 3.4 Broad historic categories of timber design (based on Yeomans 1985)

Description	Period	Characteristics
Vernacular	Prehistory to medieval times	Small-scale structures for domestic use, either entire huts or roofs. Joints would be fastened with lashes or some notches. Locally sourced poles, saplings, branches or driftwood as the structural material.
First large-scale structures	Roman and Byzantine empires	Major roofs on large-scale public buildings or private villas, mainly in early Queen posts and tie-beam trusses. Use of imported timbers. First attempts to codify and transmit building practice (Vitruvius).
Craft based	Medieval to end of 17th century	Ranging from cruck frame through crown post roofs, to the development of trusses. Great variety of crafted joints, often locked by hardwood pegs. Oak as the predominant structural timber.
First softwood structures	Mid-17th century to early 19th century	King post and Queen post trusses with some ironwork, but mainly carpentry joints. Deeper, narrower joists in floors. 'Double' floors. Imported Baltic softwoods.
Early engineered	Early 19th century to 1920s	Variations of King post trusses with some ironwork, with metal-assisted connections. 20 m spans achieved. Most modern timbers available, machine sawn, with sections placed vertically.
Modern	1920s to present	Fully calculated designs. Timber connections and a range of engineered timber available, followed by reliable structural plywoods. Codes of practice and standard designs.

stability strategies, layouts, connection details or forms that are often directly expressed in the architecture of the building, depending on local resources or the ability to trade for better quality materials.

Examples of early timber frames in vernacular environments are found in hut types tied with ropes, ranging from plain tepees to elaborate and spacious types like the lakeside Crannogs made of straight saplings around a centre pole/roof tree that supported a conical roof.

Until medieval times timber was used in roofs or embedded in rubble walls as reinforcement, but the increasing provision of good quality, long timbers meant a gradual prominence in the skeleton. Forms like cruck, post-and-beam (Fig. 3.21), box, or aisle frames were developed. The skeleton in a post-and-beam structure would require diagonal bracing within the wall planes to resist lateral forces (brickwork, infilling panels, etc.). The upper floor would cantilever at a short span, providing additional space and structural efficiency of the joists of the floors through a moment-resisting joint with the supporting post.

The more elaborate cruck frames would originate from a direct use of the form of trunks, which were split longitudinally and the two halves (cruck 'blades') would be joined at their middle with a collar (tie) beam and then at their ridge to form a stable A-frame. Even more elaborate timber

3.21
Post-and-beam frame building in Newark, England

forms are found in the aisle frames, where a central nave is flanked by two side aisles, similar to many contemporary, often stone-built churches, creating in England some impressive halls or even barns.

Contemporary medieval roofs used king or queen posts (depending on the use of one or two posts, respectively, which interestingly carry no compressive forces) and original examples using highly strong oak are still found today in Durham and Lincoln cathedrals. Further elaboration of all these achievements would eventually produce sophisticated and prominent timber enclosures such as hammerbeam roofs (Westminster Hall 1394 or a reconstruction in Stirling Castle in 1999): these depend on corbelling action to reduce the spans and transfer loads to stiffer spandrels below.

This concept had a completely different emphasis in traditional Chinese construction (Fig. 3.22). In raised beam construction, the roof skeleton consists of several levels of raised beams connected by a complicated joint called *tou-kung* (a corbel bracket). Alternatively, in *chuan-dou* construction, the beams and purlins are made of slender and short timbers, which are supported on relatively dense columnation.

In some cultures, timber is the exclusive material for entire buildings (structure and envelope) as in the case of late Ottoman Istanbul (then Constantinople). Entire areas (Sultan Ahmet, Zeyrek) were built in timber from the eighteenth century either as town mansions (*konak*, Fig. 3.17) or terrace houses, while even more refined and extensive structures are found in the seaside villas along the Bosphorus or the Princes Islands (*yalı*). The Greek Orphanage on the Pringkipos Island on the Marmara Sea (Buyukada), for example, is one of the biggest timber buildings in Europe.

A *konak* is a balloon frame, two or more storeys tall with cantilevered porches (like *erkers*). They are considered to have better seismic resistance compared to masonry because of timber's flexibility, so diagonal bracing was critical but unfortunately not consistently applied. Their safety in the case of fire, however, proved to be insufficient due to the continuous posts of the balloon frame, which would spread fire, and the vicinity of the porches that can transmit fire between buildings as happened in the fire of Pera in 1870.

3.2.3 Iron and steel frame

Steel frames owed their vocabulary initially to timber structures until robust moment-transmitting joints were fabricated. The system further evolved

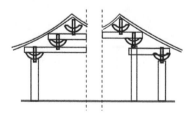

3.22
Tou-kung (or *dou-gong*) and *chuan-dou* based on corbelling principles in historic Chinese timber construction

with the wider understanding and implementation of welding, especially from the 1950s and later with the introduction of computer aided manufacturing (CAM) which enabled precisely cut and welded plates and sections. Once steelwork was liberated from solid fire-proofing, buildings would be increasingly transparent and lateral stiffness was provided either by braced frames (horizontal and in-plane bracing through ties or walls) or rigid or 'sway' frames (that resist both vertical and horizontal loads through moment-transmitting (rigid) connections). Stability is further complemented by the diaphragm role of the floors, which was usually composite with non-combustible concrete.

Early buildings in cast iron would be distinguished by tubular columns and cast 'crush box' joints, while beams had fish-bellied bottom flanges (Fig. 3.13) until these were replaced from the 1850s with wrought iron ones, formed by riveted plates (Campin 1883). The development of frames brought benefits such as wider spans, better overall control of stiffness and construction quality and, eventually, a method to 'design' a structure, i.e. to specify and refine scientifically the sizes and location of members.

Transparency would be pursued in skyscrapers or large infrastructure projects like stations or industrial sheds, while RC was the main system for medium to large private and public buildings. In the AEG Turbinenfabrik in Berlin (1909) the concept of a building inspired by machine functions and production was successfully explored by P. Behrens (the building has kept the original function since). The connections, such as the pin joint at the bottom of each frame, were fundamental for the design (especially the tapering columns that create the classical rhythm of the building and its reference to ancient temples, while disintegrating the envelope with glazing).

So far as the connections between members are concerned, bolts were used initially but for larger-scale constructions or speed hot riveting was used, as exemplified in the American skyscrapers. Welding is favoured for connections made in the fabrication workshop while bolting is the preferred site method – both procedures can be combined. Hinges were used in very large frames to accommodate thermal movement and reduce the costs of fixed connections (Gallerie des Machines 1900, Turbinenfabrik 1909). Fixed, moment-resisting connections had to be covered for their importance in structural performance but also due to the industrial fabrication which was not elegant, until confidence and advances in alloys or fire protection from intumescent paints permitted joints to be exposed. A large-scale exploration of the design of joints and even the combination of pins (on top of the columns) with robust, fixed connections between the beams was carried out in the New National Gallery in Berlin (1968), Mies van der Rohe's last work.

Full transparency had been explored earlier in the nineteenth century: prefabrication and deep truss girders enabled the fast erection of the iconic Crystal Palace (built in Hyde Park in 1850, relocated in Sydenham Hill in 1854, burnt down in 1936) and, equally spectacularly, in the various Victorian orangeries and glass houses in the UK. The Palm House in Kew

Gardens (1844–48) was a showcase of the flexibility of wrought iron, the new material of the time, creating the necessary spatial arrangement for the exotic plants hosted with an aisle-frame type structure that dissipates thrusts to a wider footprint.

These buildings, however, were vulnerable to fire, as the Crystal Palace showed, and enclosed buildings could not afford a similar transparency. De la Warr Pavilion (Bexhill-on-Sea 1934), an iconic expression of Modernism's ideals for closer contact with nature, has an innovative steel frame designed then by engineer Felix Samuely (Fig. 3.23). Precise welding replaced bolted connections and minimised steel sections but the steelwork was encased in concrete and tiled to reduce corrosion from the seashore environment.

The 38-storey Seagram building in New York by Mies van der Rohe (1958) exhibits a ground-breaking design that was produced by a steel moment-frame combined with an RC core up to floor 17 and a diagonal steel core bracing thereafter. This innovative scheme would be followed by many similar tall buildings. The ideal full transparency would be established later by recessing the steel frame completely, as in the glazing of Sydney Opera House by Peter Rice or the glass envelope of the Willis Faber Dumas offices

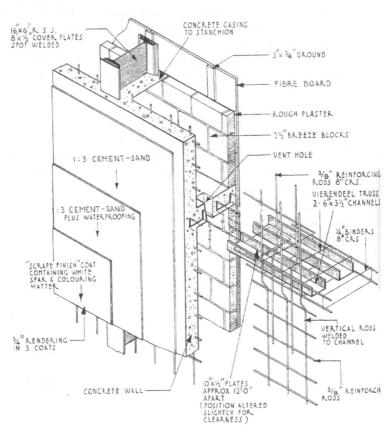

3.23
Encased steel columns in the De la Warr Pavilion (Powell and Schollar 1994)

in Ipswich in 1975, one of the most recent listed buildings in the UK in recognition of its seminal role.

Regarding the structural pathology of these buildings, fire has
been considered traditionally as a major threat. Although steel melts and
loses strength in high temperatures, very few catastrophic fires have been
recorded historically, and actually more recent high-rise buildings (like Torre
Windsor in Madrid or the Oriental Mandarin Hotel in Beijing) have been
affected. Also warfare has caused major destruction in steel structures for
example, in France during the First World War or Berlin and Dresden in the
Second World War, often in more decorative domes that were clad only in
metal sheathing.

Some recent failures include mainly historic Victorian pier
pavilions in England (Brighton West Pier 2003, Weston-super-Mare 2008,
Southend 2005 – Fig. 3.24, Fleetwood 2008, Hastings 2010). They are often
made of cast-iron columns and timber, and in some cases despite the dev-
astation the columns have not deformed greatly. The strength of the members, integrity of connections and space would, however, be compromised
beyond repair, while sometimes even the pier itself has been affected, or at
least the platform.

Systematic fire prevention and regulations in the twentieth century have managed to contain great fires. Key areas have been the spreading
of fire to other floors through long steel members so barriers are recommended. The collapse of a building is often difficult to understand: initially
the heat load is variable within an enclosed space, depending on the presence of oxygen and combustible fuel. The steel members will lose strength
and stiffness at different points according to the temperature they are
subjected to or the exposure and geometry of the member. Ductility, load

3.24
Reconstruction
of Southend pier
following the fire in
2005 (Rowson 2009
– courtesy of the
New Civil Engineer)

redistribution and eventually failure may happen in various scenarios so often a more phenomenological approach is taken to study failure empirically, from real case studies or large-scale experiments, or to apply performance-based design.

Focusing on the members, in flanged beams, when combined with concentrated loads or the impact of falling debris, thin webs in a beam would behave like a thin column and buckle or warp and distort. Buckled columns are not fit for purpose and can rarely recover their function. Cast-iron columns can suffer brittle cracks and even lose parts of the section, but eventually they can be stitched and kept in service.

Concerning the effect of other, excessive loads, connections are critical because they are fabricated and their failure depends on the process or the strength of the components rather than the steel itself. Failure of rivets and bolts is characterised by their shear or bearing performance but also the bearing capacity of the member or plate. Failure appears as tearing-out of the rivet or bolt through a plate, or of the plate across the bolt holes. Relaxation of the connection due to heat elongation is also probable.

In welded connections, failure depends on the strength of the weld (and consequently the execution) and fatigue after continual load changes (especially in wind or seismic action). Bridge bearings are often, even today, made in cast iron because of its high compressive strength and they can fail by crushing. A recent example of catastrophic failure of a joint is the Interstate 35W bridge (I-35W), Minneapolis in 2007 (Hansford 2008). Collapse of the 1964 truss bridge was attributed to a design flaw of the gusset plates that connected the girders together in the main truss, making them unable to withstand the stresses they were subjected to.

Failure of the material under rather static yet high imposed loads can seriously affect buildings without a clear structural hierarchy. Such problems can occur in nineteenth-century textile mills or warehouses characterised by an internal structure of cast-iron beams and columns, supported on and stabilised by load-bearing masonry external walls. Ties at jack-arch floors can be infrequent therefore the robustness of cast-iron columns can be problematic as not enough alternative paths exist, making disproportionate collapse very possible.

3.2.4 RC frame

The pathology of concrete frames will deal primarily with stability, one of the two key aspects, while durability of the envelope affects frames to a lesser extent when compared to shells, for example. Concrete is expected to form monolithic members and connections around the reinforcement when cast in situ, so most of the key developments examined earlier focused on the strength and appearance of the joints, and the beneficial use of the stiffness. Failure at a point is 'designed' to occur by yield of the reinforcement bars and subsequent crushing of the compressive zone, a pattern that should give enough warnings beforehand. The plastic hinges that form this way cannot carry more loads but any increase will be transferred to adjacent sections until, ultimately, a sufficient number are formed to create a mechanism and collapse.

Failure of joints was an early concern and the inspiration from the more established steel structures characterised by hierarchy between primary and secondary beams, gave way to the benefits of more even load sharing in RC. Of particular concern for early designers was how to resist punch shear around the columns, but eventually expressive solutions, such as the mushroom caps used by R. Maillart or Owen Williams, were visually out of place, even if they developed an architecture of their own in F. L. Wright's Johnson Wax offices.

The continuous reinforcement in concrete ties the entire building together, as long as the re-bars (reinforcement bars) properly overlap, and distributes loads more evenly. The scheme has to be completed by 'strong points' for lateral stability, a more critical problem in seismic zones where asymmetric inertia may cause heavy torque. Often such points are the lift shafts or stair boxes and they are combined with reinforced walls on the perimeter or cross walls, as was explored in an emblematic earlier application in Highpoint 1, London 1937 (Yeomans and Cottam 1989). Infills by brickwork can also improve the in-plane stiffness of RC frames, in addition to the robustness of the joints.

The layout of the building should also be checked for its serviceability performance, i.e. it should avoid excessive deflection or cracking which may not be detrimental to stability but cause service problems or discomfort to the occupants. In modern construction, such limits have been set as span/250 (where the deflection is noticed by user) and span/350 or 20 mm (where partitions and finishes can be damaged). Excessive deflection was a crucial problem during the refurbishment of Owen Williams' Boots D10 factory in Nottingham and had become a major historic characteristic of the (technically inefficient) design of Fallingwater that eventually threatened its survival.

Failures in general are rare and mainly occur due to lack of ductility of structure (large deformations could have been absorbed), even more critically during a fire or seismic activity. The major protection method is to provide sufficient cover of reinforcement and avoid spalling and exposure. Failure can also happen due to disproportionate collapse and lack of robustness, especially in precast buildings, as was seen earlier.

Movement of the structure can be a serious problem when an RC skeleton might not have the necessary ductility or ability to accommodate movement. Thermal expansion and long-term deformations (creep, shrinkage) require movement joints 25 mm wide at 50 m centres or spaced at 25 m for open or exposed structures in current practice. Differential settlements due to foundations or soil conditions also cause high shear forces around the joints.

Material and envelope pathology affects the presentation of RC structures, especially where the architecture was based on plain and simply textured surfaces in exposed concrete. Even in the case of rendering, low durability may cause cracks and water ingress will reach and corrode the reinforcement, causing spalling. This was the outcome of bad craftsmanship in E. Mendelsohn's Einsteinturm in Potsdam (Fig. 3.25) and insufficient reinforcement cover in B. Lubetkin's Highpoint 1, which led to the necessity for patchy repairs throughout their lifetime.

These projects were originally designed with the best intentions, within the limit of technology, but active disregard to correct practice was a problem that often characterised F. L. Wright or Le Corbusier's work, as in Villa Savoie (1929) where extensive roof leaks finally forced the owners to abandon the Villa in 1938, allowing it to physically deteriorate until 1959 when restoration was promoted by A. Malraux.

3.2.5 Arches and bridges

Masonry arches and bridges should ideally be designed to develop primarily compressive forces to make the most of the high compressive strength of the material. Otherwise, large thickness will be required to contain bending and reduce tensile forces, or reinforcement strategies should be employed. The success of such schemes depends on the containment of horizontal thrusts at the base. A minimal spread of the supports can dramatically change deformations and loads as will be seen in the case of vaults.

Due to the discrete nature of their masonry and the long times lime mortar would require to set, centring has always been necessary during construction. The design of the arch would therefore need to consider the complexities of erecting and dismantling the centring, two equally delicate processes. The location of supports and the ability to temporarily bridge a long or inaccessible span conditioned the design of bridges such as

Cracks along re-bars
Spalling
Concrete patches
Exposed brickwork

3.25
Deterioration of the vertical telescope tower in Einsteinturm in 1995 before the repairs, SW elevation (after Hoh-Slodczyk 1999)

Castigliano's Ponte Mosca in Turin (Castigliano 1879). The issue required equally creative responses in RC bridges, such as the 133 m-long Salginato-bel road bridge by R. Maillart (1930) or Kingsgate footbridge in Durham by O. Arup (1963).

Greek arches date mainly from the fourth century BC and they are defined more by construction than geometry, using carefully shaped vous-soirs to fit without mortar, as in the barrel vault at the entrance to the sta-dium in Olympia. It was the Romans, however, who understood the arch's true potential for long, durable and fireproof spans, which would enable major infrastructure projects like aqueducts and bridges, many of which still survive. Voussoir construction was also used to construct the multiple arches of aqueducts, successfully spanning 4–5 m each (Pont du Gard, Seg-ovia). Medieval bridges were built and often reconstructed in various stages, based upon massive piers to withstand thrust from the arches, water pres-sure and scour, as well as to provide a solid support for the formwork. The latter was critical where formwork was reused for economy in multi-span bridges as spans had to be almost identical in order to fit (Berwick upon Tweed, Blackfriars Bridge in London).

Most bridges up to late medieval times would follow geometric profiles for ease of construction, such as semi-circular (Puente Alcántara), stilted or pointed forms (London Bridge), in line with architectural trends. Some interesting examples can be found among the cases where heavy piers were used (Yalding Twyford bridge) or spectacular single spans like Pont du Diable, Ceret in 1341 (spanning 45.5 m), Mostar's Stari Most (1567, span 40 m) or multi-spans like Arta (c.1606, spanning 24 m, 15.8 m, 15.4 m and 15.2 m).

Heavy spandrels or hump-back profiles characterise most of these bridges in their attempts to accommodate and reduce the thrusts to the supports. Some early variations arose from an aesthetic debate, marked by the introduction of a flat elliptical profile for the first time in Santa Trínita, Florence (1569), spanning 32 m. Although it appears this was done more for navigation and aesthetic reasons (a flat profile exerts higher thrusts) it opened up the design to other geometric forms. After the mechanics of arches were studied by R. Hooke and Coulomb, the bending (and tensile) forces generated by circular profiles were understood and thrust lines would be identified. Forming the arch along such lines favoured axial compressive forces and profiles that suit physical properties of stone and brick, allowing longer spans.

A series of major bridges in masonry were built as confidence to the theory and technology was growing towards the times of the Industrial Revolution. Many of these new designs were prompted by the need to over-come scour that undermined the piers or the need for more traffic, espe-cially in the cities. In particular, developments in Britain were also pushed by the constitution of the canal network and the need for aqueducts. The new era was signalled by Westminster Bridge (1734). Later, the pioneering Pontypridd Bridge was given a unique high rise against a 43 m span and was concluded by W. Edwards in 1756 following a series of disasters that tested the (insufficient) weight of the spandrels or the rise (Ruddock 2008). Before experimentation with iron and steel, long-span masonry bridges were designed by Robert Mylne or major civil engineers such as John Smeaton, John Rennie (Lune Aqueduct 1798, 21.3 m each span; the flat arches of Esk Bridge in Musselburgh, 11.3 m each span) and Thomas Telford (the elliptical Over bridge 1828, spanning 46 m; Dean Bridge, Edinburgh 1830, spanning 27.4 m). The life and works of these and many other pioneering British civil engineers of those times are discussed in the biographical dictionary by Prof. Alec Skempton (2002). Thrust lines and catenary forms would drive the design of bridges and the theory would be established after the seminal work by A. Castigliano on Ponte Mosca over the river Dora in Turin (Fig. 3.26), based on calculations following his elasticity theory (Castigliano 1888).

Essentially a bridge of this type is a thin arch (often tapering to the crown) with loose fill (like gravel or charcoal) above its springing to resist the thrusts. Many of the eighteenth- and nineteenth-century innovations were about the profile of the arch, improvement in construction quality and balancing the stiffness and durability of the spans in order to distribute the loads more evenly (McKibbins et al. 2006). Backfill of the spandrels was important for the equilibrium (and also for the platform) and the walls were built primarily to contain the fill rather than improve the stiffness (Fig. 3.27). There was a fine balancing act, and infill that was required for the platform but was excessive for stability would be relieved with arches, like the holes in the spandrels of Pontypridd Bridge.

Maillart's treatment of spandrels in his RC bridges reflected this observation. A critical stage in the construction was the type of mortar as most bridges are close to water and need waterproofing. Non-hydraulic limes combine the necessary strength with flexibility and durability but require very long times to set, which requires careful planning of the removal of the centring. Massive viaducts have also been built in concrete which was not used as reinforced but merely for its ability to form heavy spandrels at the form of arches. See, for example, the concrete arches of Glenfinnan Viaduct (1908–11) built in Scotland by Robert 'Concrete Bob' McAlpine.

A vast number of masonry bridges have been built in countries such as the UK and they are still in use today for road and rail traffic or as aqueducts in canals. Age and increase of the loads require assessment of their strength, but also other actions have to be considered nowadays,

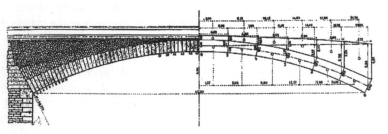

3.26
Alberto Castigliano's study on Ponte Mosca (Castigliano 1888)

3.27
Construction of the
spandrel of a masonry
bridge (Gauthay 1843)

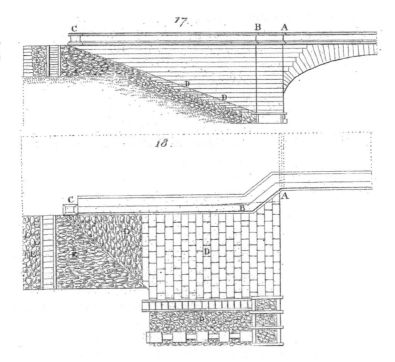

such as high point loads, impact or fire. These bridges probably have high reserves but often give few signs of imminent collapse. Because of their mixed construction system, the role of the spandrel fill especially in spreading loads to the arch barrel under critical loading combinations still needs to be better understood. The major problem, however, is the number and location of these bridges, which makes it difficult to take stock and increases the costs of inspection, therefore analysis needs to be quick and conservative.

Cracking of masonry under bending leads to the creation of hinges and four are capable of transforming an arch into a mechanism and causing collapse. Figure 3.28 (University of Salford, from McKibbins *et al.* 2006) shows the collapse pattern under a point load at third span and the

3.28
Four-hinge failure
mechanism of a
masonry arch bridge
when live loads are
simulated as point
loads, in practice and
in theory (McKibbins
et al. 2006)

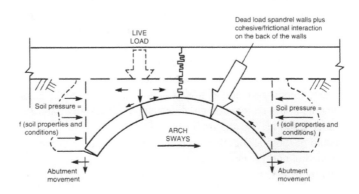

resulting sway. This simulates a live load from a stationary vehicle as a quasi-static load. The fill (and secondarily the spandrels) act as a reaction to sway but in cyclic loads from traffic (which often cause fatigue to masonry) the arch might not recover its geometry and could have permanent deformations built-in.

As critical infrastructure, masonry bridges have to fulfil certain performance criteria that deal also with serviceability, i.e. do not impede the function they serve or cause disruption elsewhere (underlying rivers or roads). Such defects are usually cracking of the arch or detachment in multi-ring arches, movement of spandrel, piers and abutments, deterioration of masonry or the parapets. Some of these problems can occur from differential settlement of the piers or disturbance of the backfill (increased loads due to water saturation or instability due to loss).

As mentioned, the assessment of the condition by in-situ inspections is difficult and methods have to be rapid and easy to handle. Examples will be discussed in the next chapter but desktop study is usually facilitated by the existence of many original data or earlier condition surveys, as they are relatively modern structures.

Analytical methods are always being developed and are divided between semi-empirical methods, limit-state analysis or continuous mechanics methods (McKibbins *et al.* 2006: 139). MEXE (Military Engineering Experimental Establishment) is an empirical method based on tests that evaluates the critical provisional axle load (PAL) from the performance of a 'standard' arch barrel. It is a quick, well-implemented and conservative tool, most suitable for very stiff or irregular arches. Limit-state analysis calculates the load required for a thrust line to 'move out' of the body of an arch forming enough hinges to cause a mechanism (an example was discussed in Chapter 2). Finally, solid mechanics methods have been applied, such as Castigliano's non-linear analysis and its 3D exploration with the FE method, which simulates load redistribution following plastic failure, or Discrete Element simulation of discontinuities between the blocks.

3.2.6 Mass vaulted structures

Similar criteria apply in the construction and performance of solid barrel vaults, with the difference that there is no need for a flat access platform and they are often roofed (Rosslyn Chapel, Edinburgh), except in rather hot climates where they are exposed (Seville) and often provide the necessary thermal mass for comfort (traditional houses in the Cyclades in Greece).

Referring always to solid structures, the transition from plain barrel vaults to complex Gothic cross vaults is characterised by geometric requirements in the distribution and hierarchy of space, the use of materials in large amounts and the effect of the containment of thrusts in the architecture of the building. Similar to arches, adjustments in the stiffness of the spandrel and thinning down the section of the crown could distribute loads more evenly. Mass vaults, however, have a thickness that can accommodate many actions and their corresponding thrust lines, therefore the redundancies are high.

Vaults that survive are mostly made of stone blocks as well as a few plain adobe ones, as some vernacular barrel vaults are still built in Egypt or Iran (although not true vaults as they essentially follow a tilted corbel pattern). Roman shells are the major type to survive from antiquity and a very fertile exploration produced domes (Pantheon), barrel vaults (Basilica of Maxentius and Trier), cross vaults (Diocletian Baths), cloister vaults (Domus Aurea) and segmental vaults (Temple of Venus and Rome).

Single curvature (barrel) or synclastic vaults (domes) had continuous supports along their perimeter, which simplified construction and performance. More geometrically complex vaults generated the richer spaces of the Imperial Roman period and could provide larger openings for daylight. As a result, their fabric springs from concentrated points set on pendentives: conoid forms that could be built horizontally or cast and, due to their stiffness, reduce the true span. Interesting late Roman structures such as the 'Minerva Medica' (Horti Liciniani) introduced the gradual spread of thrusts to smaller perimetrically arranged vaults, which became a prototype for medieval structures (Fig. 3.29).

Their construction in Roman lightweight concrete was often assisted by the ribs (*ducti*) in brick tiles incorporated in the masonry (Giuliani 1992). Tiles can also serve as a permanent formwork at the intrados. The critical thrusts can be reduced by thinning the shell towards the crown and using lightweight materials or amphorae. For example, the Pantheon dome has a 50 m diameter and its thickness reduces from 5.9 m at the base to 1.5 m at the oculus. The barrel vaults in the Basilica of Maxentius span about 28 m, with an 11 m rise and 18 m length. Thickness increases from 2 m at the apse to 4 m at the atrium.

3.29
Embedded ribs at the so-called 'Temple of Minerva Medica', Rome

Byzantine vaults can be considered as part of the same tradition but they are characterised by less complexity and ambition. Domes and barrel vaults are almost exclusively used to roof churches, defining a centralised space that enhanced experiential participation. Especially in the smaller churches of mainland Greece, there is a hierarchy between domes and their squinches (pendentives) that shed their weight and thrusts to the walls and the tri-partite apse. Masonry is often in brick, with thick mortar joints and is characterised by lack of accuracy in the execution or planning (Ousterhout 1999), which however became more precise after direct contact with Western builders following the Fall of Constantinople in 1204.

Technical refinements are more evident in the Gothic treatment of stone vaulting. The geometry of such forms initiated from the limitations of semi-circular arches in creating flexible, non-square footprints and spaces. If the same radius had to be used to standardise the voussoirs then pointed arches were produced. Ribs, as in Roman times, were used as a construction device to set up the centring and the guides that materialise the curvature of the vaults (Fitchen 1961). Function as strengthening elements although effective on structural performance is truly a post-rationalisation. Most probably the masons had an intuitive understanding of thrusts and how they affect stability, as analysis shows the use of flying buttresses was a conscious design choice .

Successful experimentation with most of these elements started in Durham choir (1093–1133) and later in the apse of St Denis in France (1130–40), the two seminal buildings where construction advancements were pushed by the functional and theological need to bring natural daylight inside the building. The possible limits of these principles were experimented with in the skewed vaults of Lincoln Cathedral (1192–1265). Their innovation, enabled by the elegant use of ribs, explores the potential of boldly detaching form and elements from their function (Kidson 1986) and the generation of unorthodox new surfaces that do not result from construction practice.

These examples frame the technical exploration of structural schemes and vault design in a period that ended with the failure of Beauvais Cathedral in 1284 that reached the limits of slenderness. The overall scheme of a Gothic building clearly highlights the dynamic equilibrium between the thrusts of the vaults and the counteraction from the buttresses (Fig. 3.30). Lack of balance will result in visible bow of the lateral wall, transfer of the loads from the nave with an offset and serious distortion and collapse. Major deformations are observed today in the choir of Beauvais and in Vitoria Cathedral.

Focusing on the structural performance of the vaults, structural analysis (experimental and numerical) has shown high strength reserves under dead load (Theodossopoulos *et al.* 2003). It is crucial, however, that thrusts are contained, as even small movements of the walls out of plumb can cause high deformations. Such weakness can result from creep deformation of the buttressing masonry or insufficient balance of the loads of the

3.30
Deformation of
the nave of Burgos
Cathedral under self
weight. FE model
magnified 1,000
times

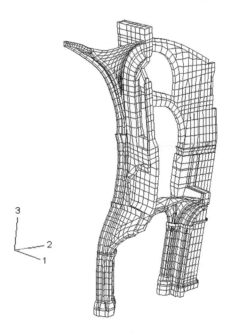

vaults. Some aspects of service conditions such as typical longitudinal cracks at the high vaults are illustrated in the FE analysis of Durham Cathedral (Fig. 3.31) and Table 3.5 presents lateral displacement, *u*, in heavily deformed churches, suggesting a serviceability limit of spread equal to 1/100 of the transverse span, *s*.

The vaults deform by the spread of their supports (outwards at the upper portion, inwards at the aisles) ultimately failing through a combination of a hinge line along the intrados followed shortly afterwards by a transverse detachment at the wall edge and abutments. This is a progressive failure that is further accelerated by the instability of the elevation due to its slenderness.

Structural interpretation of the origin and propagation of failure till collapse using structural analysis of major Early English cathedrals with a biaxial failure criterion for the stone masonry (Fig. 3.6) established the values in Table 3.6 that can serve as a reference (Theodossopoulos 2008).

3.31
Crack pattern from
the FE model of
Durham Cathedral
choir vaults
at a supports
spread of 330 mm
(Theodossopoulos
2008)

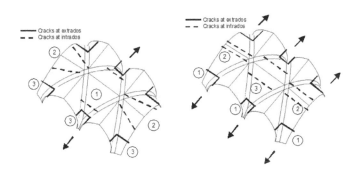

Table 3.5 Serviceability limits in Gothic vaults

Case	Location	Spread u (mm)	Span s (m)	u/s	Source
Vitoria Cathedral	Nave, bay 4	160	9.24	1/58	[Azkarate 2001]
id	Aisle, bay 4	200	5.5	1/28	id
Beverley Minster	Aisle, S transept	—		1/19	[Bilson 1906]
Blyth Church, Notts	Aisle	—		1/13	id
Kloster Maulbronn	Dormitory	170	4.50	1/26	[Jagfeld 2000]
León Cathedral	Nave	150	7.5	1/50	[Martínez et al. 2000]
Holyrood Abbey	N aisle	75	3.77	1/50	[Mylne 1766]
Holyrood model	S aisle	30	0.96	1/33	[Theodossopoulos 2003]
Holyrood FE model	S aisle	96	3.77	1/40	id
S. Angelo vault model	Nave	260	7.36	1/28	[Ceradini 1996]

Table 3.6 Spread of the supports of a Gothic vault at failure as assessed by FE modelling (Theodossopoulos 2008)

Case	Location	Spread, u (mm)	Span, s (m)	u/s
Durham Cathedral	Nave	330	11.22	1/34
Canterbury Cathedral	Choir	670	13.1	1/20
Wells Cathedral	Choir	275	9.64	1/36
Lincoln Cathedral	Nave, tierceron	503	12.5	1/25

Failure can happen due to a wide range of actions in their long lifetime. Design errors caused the partial collapse of the extremely slender piers in Beauvais Cathedral in November 1284 and the excessive deformations that can still be observed today set the limits in the proportions a Gothic church could reach. The configuration of the *porte-a-faux* junction of the upper structure towards the aisles, or the location of the flying buttresses are considered as the reasons. Vitoria Cathedral illustrates another design problem, i.e. the addition of flying buttresses after lateral deformations appeared: even such a remedy was not enough and arches were built inside the nave under the triforium.

The low strength in lateral movement is essentially the problem behind collapse due to other causes, such as earthquakes. The collapse of two vaults at St Francis Basilica, Assisi in September 1997 occurred as excessive volumes of the loose spandrel fill were able to shift during the acceleration and load the shell non-uniformly, increasing the tensile stress close to its vertices and causing non-recoverable curvature inversions. The innovative repair is discussed later (Section 6.8). Similar damage occurred after the 1980 Irpinia earthquake in the Cathedral of Sant'Angelo dei Lombardi. Other significant activities include sharp changes in the water table (York Minster) or bomb shelling, as in Reims, Soissons (Fig. 3.32) and Noyon cathedrals during warfare in 1915. The localised damage in the vaults of

3.32
Damage of the vaults
in Soissons Cathedral
after bombardment in
the First World War
(Gilman 1920; Arsène
1918)

Reims shows that ribs incorporated within the masonry provide relatively substantial stiffness at adjacent areas.

Other aspects of pathology in service conditions are usually caused by dampness in the spandrel fill. This area often fills up with debris from repairs in the roof or lack of maintenance and often attracts water infiltrating from the roof, which cannot drain. Due to limited access, it is an area often neglected and it is typical in vaults for there to be damp in the interior of their spandrels. Construction distortions are also a typical problem of a geometrically well-defined form like a cross vault that cannot fit on the top of piers already distorted, especially during long periods of construction. Often the behaviour and deformation of the church would have somehow stabilised and the redundancies of the masonry combined with the stiffness of the spandrels are able to accommodate the differential thrusts that are expected to occur in these conditions.

3.2.7 The designed shells

In contrast to the previous type, the shells that are considered to have had their sizes and form designed using modern methods have much less redundancies but are more carefully planned and constructed. Most of these forms also follow technology that is still relevant to us and can be applied with some modifications. Chronological distance, therefore, is not a single criterion for consideration as a historic structure and they offer the opportunity to reflect on what exactly constitutes the study of the history of construction today. Early shells are a case of creative renewal of a traditional form and the study of this act of innovation represents the major steps of the construction industry in the twentieth century and also its limitations.

Some of the best-known designers of the twentieth century, such as the Guastavino Company, E. Dieste, or H. Isler should not be included in this discussion because their technology is still relevant today and many current designers are still directly influenced by their work. Technology has also moved to other systems, such as prefabrication (Duxford Air Museum), composites (the work of M. Eekhout) or sculptural forms completely liberated from the function (see the bus station at Casar de Cáceres). They all recognise the plastic qualities of the solid spatial envelope that creates open, non-directional spaces that evolved from the earlier ribbed constructions to very smooth mathematical or free forms.

The history of shells moves between periods of technical creativity and their demise as mainstream structures in the 1970s. Pioneers such as F. Dischinger (1887–1953) first saw shells as mathematical forms, expressed in his patented thin RC semi-spherical dome for the Zeiss Planetarium in Jena. The work of E. Freyssinet in the field is characterised by the achievement of span, volume and enclosure, as materialised in the impressive Orly hangars in 1913. They were formed by deep parabolic RC arches on the outer surface (86 m span, 50 m rise) that hint to the catenary of loads, connected laterally by thin slabs.

The mathematical theory became complex, especially in the area of supports and simplified versions required certain conditions to fulfil, such as ribs along intersections and edges or diaphragms that limited the spatial use. Various forms of ribbed shells attempted to incorporate and hide the ribs (I. Sanchez del Rio – see Fig. 3.33) but E. Torroja managed to create and establish proper smooth shells by working on the principles of the theory. His Algeciras Market (1933) is a thin RC shell with 46.82 m span at a 44.10 m radius. It is interesting to make a comparison with the thickness of the vault of the Pantheon: the shell in Algeciras is much thinner, 90 mm at the crown, and was built in eight spherical segments, where the necessary diaphragms were incorporated and supported on columns at the perimeter.

Many barrel vaults were built as industrial structures after the Second World War since they could achieve large spans in a fireproof construction that also let sufficient daylight in. The limitations from diaphragms and edge beams eventually set a specific form for such uses and it was more poetic explorations, as in the work of E. Saarinen (TWA Terminal, Kresge Auditorium), that inspired designers. Among many other innovations it is worth noting prefabrication and standardisation as explored in the ribbed RC shells by P. L. Nervi in his expressive Orvieto Hangars (1935, 44.8 m span) or Palazzetto dello Sport (1957, 61 m diameter).

3.33
The ribbed RC shells of the Fourth Water Reservoir in Oviedo by I. Sanchez del Rio (1928)

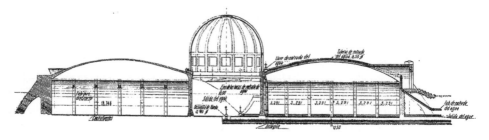

Attempts to further simplify the solution focused on the reinforcement (to avoid the excessive density around the edges and in areas of curvature change), the analytical process (split between the straightforward membrane solution and complex moment regimes in the supports) and mathematical constructs to help engineers capture the complex distribution of shear and moments in the direct design of the reinforcement (Sutherland *et al.* 2001).

Despite continuous innovation, the theory and construction of shells, especially the temporary works and formwork, became very complex, which increased costs especially for engineers or contractors who had no building experience on these forms. Inevitably in hard financial times such as during the 1970s petrol crisis, expensive structures with a 'heavy' critical path, such as shells, are not feasible and the period marked the decline of such forms.

Their pathology shows certainly most of the signs of a typical RC structure. Reinforcement exposure due to the low durability of concrete is often the key problem, which is exacerbated by the typically thin cover of early RC structures or the dense reinforcement. Usually concrete would be painted rather than rendered, in order to mask the lack of control in quality (and differences in subtle textures) and discoloration due to time. The containment of thrusts is crucial and more efficient solutions compared to historic shells are provided, such as incorporated ties (sometimes even underground), edge beams or more robust supports with wide footings anchored with piles. Water infiltration can also be a problem where the shell is exposed for aesthetic reasons and because of the span, large thermal movement has to be accommodated, often with pin joints in the supports.

Once the support conditions are guaranteed, the shell itself is a very strong structure. There are significant cases of survival under very adverse conditions, such as warfare. Many of the Spanish pioneering shells were damaged during the Civil War: Sanchez del Rio's Deposito de Aguas in Oviedo (1928) or Torroja's Zarzuela Grandstand were pierced by bombs and, similar to Gothic vaults, there were sufficient redundancies to allow loads to follow alternative paths. The collapse of Torroja's Fronton Recoletos in Madrid in 1942 was the result of the vibrations the structure suffered earlier during bombardments in 1938.

3.2.8 Drystone construction

Among the various other types that can be examined, there are interesting technical and conservation issues with a less-known group of prehistoric or vernacular buildings based on drystone construction. In essence such structures are characterised as low-tech masonry construction where material resources or practice needed for solid bonded masonry was not available, especially lime for mortars or timber for extensive scaffolding.

There is a wide variety of forms in Europe from all periods until even the eighteenth century. Prehistoric chambered cairns are found in Scotland (Neolithic, 4000 to 2400 BC) or Almería (Tholoi at Los Millares, Neolithic to Bronze Age, 3200–2200 BC), while the Mycenaean buried tholoi (around 1250 BC) are the most developed construction form. Towers have also been built, like the Bronze Age Nuraghe (eighteenth to fifteenth centuries BC) in Sardinia, the Maltese Girna and later the Balearic Talaiots and the Scottish brochs in the late Iron Age period. Drystone construction continued in the Gallarus Oratory in County Kerry, Ireland, an early Christian church in corbelled construction (dated with much uncertainty between the sixth or ninth centuries AD), up to the Scottish Blackhouses or the Apulian Trulli in the sixteenth to eighteenth centuries. Drystone construction is still used nowadays in retaining walls and experimental research has shown that strength against bulging is influenced not only by the wall profile and area of the footprint but also build quality, presence of voids, and shape and interlocking of the blocks (Mundell *et al.* 2010).

Corbelled construction was essential in the creation of domes (cairns, tholoi, trulli) and equilibrium was often secured by increasing backfill or spandrels at the base of the structure. While only 2–3 m diameters could be achieved in cairns, the Mycenaean tholoi reached a diameter of 14.5 m (Treasury of Atreus).

The study of brochs in north Scotland offers insight into the behaviour of the tower typology. They are clearly the largest and best built among them and they are believed to be iconic, probably residential, Iron Age structures developed between 250 BC and 250 AD. There is little knowledge about original space configuration and typology. Their structural scheme is characterised by an external tapering profile that sheds loads to a wider base, while the internal wall is tubular. The blocks are stabilised through gravity and friction, and they are relatively flat in the best-built and probably most stable brochs. There is no outward corbelling and the walls are further braced by ring action.

The key geometric and technical characteristics are primarily a 'bottle-neck' profile rather than straight tapering, which is probably a feature that eased construction. Many brochs are distinguished in their bases between a gallery (like Gurness) or solid base (Dun Telve). Their envelope is solid as the only outer opening is a lower door. We can only hypothesise about the internal arrangement and living conditions, and recessed timber structures probably formed the roof and internal platforms. There is debate on the role of through stones within the gallery but they probably were not wall-ties as their bond within the masonry is shallow (Barber 2009). Further voids are embedded in the base of the walls like guard-cells and their corbelled form is effectively stabilised by the weight of the upper structure.

One of the most impressive features of these complex structures is the internal staircase within the gallery. Mousa in Shetland is the most complete broch that includes this feature and it often gives the impression of the most representative structural scheme. Archaeological analysis considers Dun Telve, in Glenelg, Sutherland as the one with the most canonical proportions (Fig. 1.10). Its external and internal diameters are 15.4 m and 10.4 m, respectively, and currently it reaches a height of 11.2 m. The stacked openings that can be observed are also not well understood but plates transverse to the stack do not correspond to levels or landings as they are too close.

Apart from Mousa, all other brochs are in various states of collapse and their remoteness (mostly close to the Highlands coastline) complicates their conservation. Experimental study (Sutherland and Thew 2010) showed these structures are very resistant to ground settlement and, consequently, to any major mechanical action that could affect them. Their collapse is probably due to failure of some elements (possibly the roof or platform) that could have triggered a local damage, which combined with a later abandonment of the building and was accelerated by natural degradation and stone robbing (spolia).

3.2.9 Foundations

This category actually relates to a structural typology rather than a historic form and the technical issues are no different from those on modern buildings. A short reference is made to some historic forms, such as monumental and artificial platforms, underground buildings and substructures.

Historic foundations can often be classified in modern terms, for example, deep and shallow, and in broader terms with relation to the extent they follow the footprint of the building, such as pad or discontinuous, strip or platform (Giuliani 1992). Their function was to create a reliable surface to support the base of the building, but also to anchor it to the soil or to stronger and deeper strata. Trenches filled with rubble would be the most straightforward solution for plain buildings, while where better resources were available the rubble would be carefully compacted in layers and filled with rammed earth.

Foundations were also used in buildings whose base was formed with squared blocks but often in such cases a strip foundation of similar blocks would be built until bedrock was reached and irregular voids would be filled with rubble. This is often the case in temples (see Apollo Epikourios in Papadopoulos 2010) or large churches (Vitoria Cathedral, Fig. 3.34) but addressing the critical role of foundations was not always possible (as in the infill underneath the choir in Lincoln Cathedral) or well understood (like the 4-foot shallow strips in Salisbury Cathedral, which were almost certainly dictated by the high water table).

Adjusting the terrain to the architectural and landscape programme became more iconic in emblematic constructions but some more spectacular examples are found in castles and fortifications, as early as the Iron Age hillforts or Bronze Age palaces (like Knossos or Mycenae). The earlier attempts, as in the case of brochs and drystone structures, were

3.34
Foundations of Vitoria
Cathedral (courtesy
Fundacion Catedral
Santa María and
Quintas Fotógrafos)

the result of communal efforts and later improvements in stability and tectonics witness social organisation and maybe even structures that permitted
establishment and transmission of good practice.

Many examples of good-quality platforms are found as supports
for major classical temples, which often extend a natural plateau where
earlier phases of the temple were established. In Parthenon or Delphi, voids
were filled with rubble and contained by buttressed or polygonal walls.
Major Hellenistic or Roman buildings, however, where grand gestures were
important, were placed on massive and solid substructures that became
possible from use of concrete which cleverly incorporated vaulted voids and
large diaphragms, as in the case of the Temple of Venus and Rome, which
extended beyond the natural bedrock of Velia Hill (González-Longo and Theodossopoulos 2009). A similar and more expressive application is found in
the monumental platforms and ramps in the Sanctuary of Fortuna Primigenia
in Prenestae (Palestina). The monumentality of such platforms kept inspiring
even later monuments such as the Superga Basilica in Turin (1730) or the
Walhalla outside Regensburg (1842).

Underground infrastructure is one of the most impressive and
less-known areas of historic technological achievements. Some features
in classical architecture are inspiring, such as Eupalinos' precision in the
excavation of his rock-hewn tunnel in Samos, or the continuous function
of the Etruscan and Roman Cloaca Maxima drain in the city of Rome. Tunnels and drains were fundamental for urban modernisation in the nineteenth century (Fig. 3.35), even becoming cultural icons, for example the
first Thames Tunnel in London by the Brunels (1825–43), or the Parisian
Catacombes and *Egouts*. Subways were other major works, for example

3.35
Part of the oval cross-section of a railway tunnel to resist and balance the earth and water pressure (Lace Market, Nottingham)

the London Underground (1863), Istanbul's Tünel (1875), Budapest Metro (1896), Paris Métro (1900) or New York Subway (1904). Materials and bond are important not only for the stability but also the important waterproofing of the shaft.

Substantial supporting structures are found in important buildings that expand well beyond the natural features. Most of the Palatine or the Capitol Hills in Rome have expanded this way and their impressive substructures (*substructio*) include galleries and architectural features that combined with the function of buttressing, as in Domus Tiberiana or the Tabularium, respectively. Niches that can be carved in massive retaining walls often give a more monumental aspect to the building, while the flanking returns provide the necessary depth, working as buttresses.

Castles are projects that often combine the functionality of the defence strategy with ambitious architectural plans or even residences. They are characterised by forms in between all the previous types. Heavy defensive walls often include underground spaces, with vaulting providing strength and making best use of the stabilising effect of uniform earth pressures. Fortified cliffs are the most impressive structures in the castle typology, designed in order to emphasise control over critical strongholds and passages (Palamidi castle in Nafplio, the city walls of Ávila, Hadrian's Wall in Northumberland) or in networks of castles (like the Gulf of Patras or Bellinzona). The layout of a castle is often organised in a way to minimise the effects from the impact of artillery or direct attack, especially after the advent of gunpowder in the fourteenth century.

Underground buildings such as cellars and storage are found in castles or as undercrofts in extensive country mansions like the cryptoporticos in Hadrian's Villa in Tivoli. They are often cut and cover structures that form extensive underground networks and their usual problems are concerned with durability and failure of the waterproofing or drainage structure. Insufficient containment of the earth pressures can often cause partial collapses that render other spaces more inaccessible rather than unstable.

Chapter 4

Knowledge from structural assessment

The theory and practice presented in this book is aimed at discussing the process of major structural interventions to historic buildings. The choice of techniques and design options, which is the particular focus, is also conditioned by the knowledge of the condition of the fabric. The discussion moves now to more operative stages and the wide range of assessment methods is focused in those aspects that directly support informed design choices on the extent of intervention. A more detailed outline of the main methods in their technical context can be found in Beckmann and Bowles (2004), Forsyth (2007b), Andrews (2003), Docci and Maestri (2009), IStructE (2008) and Hollis (2005).

4.1 Aims and methodology

Knowledge of the character, structural scheme and condition of a historic building before an intervention depends primarily on the nature of the project. Desktop study on the history of the building can reveal important data on the originality of certain elements and is often a statutory requirement. A full structural assessment provides vital data on the 'health' of the structure as well as its durability. In addition, many owners or public funders require a full survey of the condition of a building in order to assess its financial value as an asset.

However collecting this information can be a costly and time-consuming exercise and often parts of the building might be inaccessible or historic records are incomplete. In many cases, however, the information required is very specific or related to certain areas and elements only. Therefore, the assessment has to be carefully planned alongside the project and within the budget available.

The key question is what information is required. In most of the projects reviewed in the case studies, it is structural adequacy (stability, remaining safety, serviceability) that needs to be known by the engineers who design the new structure or those who have to certify the intervention. As mentioned earlier, many buildings do not have an imminent safety problem but the intervention is driven by a particular new use (conversion) or specific condition (compromised architectural integrity). In those cases, the survey and appraisal is there to record the current architectural and structural condition in order to assess its character. This is particularly important information for the design process, as there might be specific aims that are not always compatible with the apparent character of the monument. There are also cases when historically significant parts might have to be destroyed and the survey is required to provide a full record.

The structural and geometric appraisal should eventually provide a clear picture of the condition of the fabric so that resources, repairs, structural layout, etc. can be planned. This is given various definitions according to the aims of the project; often depending on whether the current state is surveyed for the purpose of maintenance or for planning a major alteration.

Another way to define appraisals is to distinguish between qualitative and quantitative surveys. Although the criteria and experience of the surveyor are important, the former type of process is now quite well established and is often simplified by the use of forms and standard types of reports. It is always, however, the interpretation of damage (cracks, dampness, deformations), its scale and inter-relations with other damaged areas that need the analytical capacities of the expert who conducts the survey.

Quantitative appraisal can provide information on the value of displacements, the curvature or out-of-plumb of a wall, the deformation of a window, the amount of humidity, the level of daylight, etc. and is required in types of analysis that can handle data with such accuracy. As always, the value of such data and the ability to handle them depends on the scope of the project.

Many techniques that assess the integrity of a fabric and its continuing safety use instrumentation that is classified as invasive or non-destructive. The decision is often straightforward in modern projects of infrastructure and it depends on costs, extent of the survey, accessibility, timescale and the problem itself. Historic fabric, however, carries further implications and the use of infrared, collection of samples, insertion of jacks, etc. may be prohibited and alternatives should be sought.

Structural stability, as mentioned, is the key aspect that affects most of the projects discussed later and is directly related to the (remaining) degree of safety of the building. Similar to that, the reasonable probability of failure occurring in the project, and the costs incurred to guarantee safety against all actions in the course of its entire life cycle, have to be balanced. As a guide towards reasonable and meaningful appraisals, these are a few axioms from practice (Beckmann 2004; Hume 1997; Forsyth 2007b):

- The building is there and it stands.
- The sizes of its elements can then be verified, but not dimensioned, and their strength can be assessed in situ.
- Compared with modern schemes, the structure might seem unorthodox and several load-bearing assemblies or oversized members exist that provide many alternative load paths and redundancy.
- If calculations indicate lack of safety, they are either based on wrong assumptions or give only a partial aspect of the behaviour.

The following sections will review briefly the range of survey and appraisal techniques. As mentioned before, each technique is suitable for the type of information required, the budget or type of monument, use, how critical the intervention is, etc. The range and magnitude of data of each method as well as the type of error inherent to each method will be reviewed alongside the implications of the use.

4.2 Empirical and desktop methods

Archive research can provide, principally, data on the original construction of the building or subsequent alterations. How far the documentation is complete depends on the period of the building, as according to the process of design a full set of drawings might have been deposited to a controlling authority. State archives are the first source for pre-modern structures while city archives are expected to contain full projects submitted for building warrant or planning permission. Such archives should exist in any major city but access or organisation is variable, often depending on cultural attitudes to information. There are also specialist archives that collect information on historic buildings, such as the National Monuments Record (England and Scotland) or Monuments Nationaux in France.

In many cases, however, essential information needs to be collated from various archival sources, depending on the ownership of the building, the original commission or the way later repairs took place. An example of information available in archives is the collapse of the Abbey Church of Holyrood, Edinburgh in 1768 following an incorrect repair of the decayed roof, which was fully minuted by the owners. The Lords of the Exchequer who used to manage the Palace of Holyroodhouse at the time ordered the repair in 1758, paid the contractors the full balance on completion in 1760, ordered the closure of the church after heavy deformation in 1766 and finally asked for a report of the damage after the collapse on 2 December 1768. These reports, letters and estimates are found in the Exchequer Court (Treasury) records (Minutes, Letters, Petitions, Orders), deposited in the National Archives of Scotland (not in the Monuments Record), while a copy of the contract for the works must exist among the Register of Deeds that recorded legal agreements at the time.

The research can become more complex, as in the case of some churches in Rome where part of their site was expropriated by the state when Rome became capital of the new Italian state in 1870. Part of their history is safeguarded in their own archive, ownership may be with the Home Office and repairs are carried out by the Ministry of Culture. Information on more modern buildings might also be part of the archive of a company or an architectural practice that does not exist anymore and whose archive has been dispersed (if not destroyed).

Archive research can be a long and scientific process that might not suit the budget and dynamic of a project. In many cases, this research is part of a desktop study where more directly accessible original information can yield sufficient information that is further interpreted by the engineer or the surveyor. This usually includes recent historic maps, property titles from Land Registers, insurance reports and mining records. The desktop study also collects information on soil condition as a preparation for the site investigation and, together with the above sources, local knowledge can be of use regarding the composition and stability of the soil.

There are also quantities that do not directly result from the building but affect its behaviour, such as the amount of applied loads (imposed loads, fire load, etc.). When regulations do not cover them, experimental work can be handy but load tests should rarely be performed on the building itself.

Once sufficient background information exists it has to be combined with in-situ surveys of the state of the building in order to assess the condition of the building or establish its material or architectural character. There are various definitions of these surveys according to the information they yield or their legal role, and some of the British definitions were summarised by IStructE (2008). A Building Survey is the general term for the professional assessment of the construction and condition of the fabric of a building. More specific, a Schedule of Condition records the condition at a particular time (at the beginning of a lease or construction works) while a Schedule of Dilapidations records the necessary repairs at the end of a lease or a specific period (see, for example, Table 4.1). An Investigation Prior to Alteration is also a type of technical report required before alteration works, sometimes involving destructive investigation or tests.

Many major buildings have established a regular schedule of inspections, like the Quinquennial Reports that most cathedrals in England produce (for example Table 4.1 from Lincoln 1995). Information at a similar level is often found in Spain at the Plan Director that major cathedrals prepare (see, for example, Azkarate *et al.* 2001 for Vitoria), but this captures information only at a specific period and when planning for a series of major interventions. As J. L. Taupin had mentioned for Beauvais (1993), such documents have become essential nowadays because congregations do not frequent all areas in such complex buildings, such as the galleries, which are often prone to structural deformations. Alternatively, in some specific

Table 4.1 Example of the Quinquennial Condition Survey of a triforium gallery from Lincoln Cathedral (1995)

Bay 9E	807	East wall acceptable. A settlement crack running down the south side of the door has been recently pointed and shows no sign of movement.
	808	South wall in fair condition, the surface generally attacked by water washing, is grouted and repointed.
	809	The Arcade is acceptable. There is slight horizontal movement in the joints at low level on the eastern side of the central pier. Piers and tracery acceptable. The Purbeck on the south side is not as badly worn as that on the north but still suffering from significant surface loss.
	810	Vaults generally acceptable, a little rough on the eastern side. The vault is some 33 mm away from the south wall.

(e.g. Dormentbau Kloster Maulbronn) or emblematic cases (Beauvais, Santa Maria del Mar), surveys and research are extensive and well published (Pörtner 1992; Roca *et al.* 2010; Taupin 1993; Wenzel 2007; Barthel *et al.* 2006) and can serve as guidance for other less studied cases.

Specialist reports cover any particular field that may require the opinion of an expert, such as the conservation of a type of stone or timber, or the selection of new repair materials, the character appraisal of a complex site (especially with strong modern connections), the development of a conservation plan, etc. The latter is also a preliminary or integral part of a Feasibility Study, which helps an owner decide upon the available options for a conversion.

There is also another range of studies that are required for every type of buildings, whether historic or new. These cover preliminary aspects such as Health & Safety during construction, life cycle costing or sustainability during the design stage, or post-occupancy aspects like accessibility or workplace comfort. Certainly such aspects become even more complicated in historic buildings that never had to consider them previously and, when conflicts with the regulations or current policies occur, a negotiation for a 'relaxation' may be necessary between the architect and the authorities.

All these surveys and studies provide information about the condition of the building at a macro-scale but detailed knowledge is always required about the location of the new structure, the sizing of members (both for structural and constructional reasons), the provision of services and envelope, the planning of delicate repairs, etc. Measured surveys are tools to both record the geometry of the building and analyse its condition, through the act of transferring the data in plans or models or completing missing information (Fig. 4.1). In architecture that follows geometric rules, such as the classical orders or a Gothic church (elevations, plans or vaults),

4.1
Direct measurement of the plan of Blyth Priory Church, Nottinghamshire, using triangulations

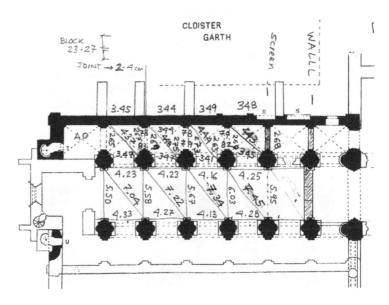

knowledge of the rules that made them possible makes the hypotheses for missing heights, incomplete elements or spacing etc. more solid. More interestingly, in the case of heavily deformed elements, application of an informed hypothesis can produce valuable observations on the reasons behind the problem, as occurred in the case of the vaults in Santa Maria la Vieja Cathedral of Vitoria, Spain (Azkarate *et al.* 2001).

There should be a balance between how many data can be directly retrieved in a survey and how far the rest can be established analytically. Depending on the aims of the project, the survey can range between a contour (which gives the outline and main deformation) and a very detailed record of every block in a mixed masonry matrix (where relations between the historic phases need to be identified). The instrumentation has changed quite dramatically from the classic direct methods. Theodolites and levels using trigonometry have been fully digitised now, allowing data to be immediately stored, transformed into real coordinates and rectified from measurement errors. The coordinates can be referred to a local system or GPS for larger, topographic scale. Ultimately, the fully automated 3D laser scanner techniques almost remove even the operator from the act of data taking, as the ability of the laser beam to create a cloud of points to any desired density offers the most realistic representation of the object scanned (Fig. 4.2).

The dense coordinates obtained from the laser scan can then be simplified to those areas of interest and traced to create the geometry of elements. Plans, elevations and sections can then be selected at every level as long as the fabric is visible from the laser scan. The data can even be exported to numerical structural models and a growing number of FE analysis programs have routines that make distinctions between solids and voids in the geometry. There is more versatility of the output as the coordinates can be transformed into a rendered model that is then imported by 3D printers to create solid models of the prototype.

The versatility of the method, however, can have drawbacks on the quality of the analysis. The vast quantity of data produced in the first instance might not be necessary and can be time-consuming to process. Above all, however, it removes the essential step of analysis, at least for those less experienced operators who think that the geometric representation of the prototype is served only by the point of clouds. Practice with essential techniques like triangulations in a plan, assessment of a wall thickness, establishment of levels, etc. offer the surveyor direct control and, even, the confidence to identify the exact type of information required. There are

4.2
Information from a 3D laser scan survey of the collapse of a model broch (Sutherland and Thew 2010 – courtesy AOC Archaeology)

several other techniques, each requiring a specific process and with certain ranges of error and they are more extensively discussed in specialist publications (Andrews 2003; Docci and Maestri 2009).

Returning to techniques that provide large quantities of data, photogrammetry and its simpler version of rectified photography have combined the benefits of realistic representation with its transition to linear drawings that extract the essential information. Even this process has been largely digitised nowadays with software that rasterises the image. IT allows now a very wide integration of almost every medium: photos or surface textures can be restituted and pasted into a 3D scan survey and features from the photo can be extracted as layers from a photographic image. Conversely, data from carefully taken photos can inform a 3D survey and complete missing or inaccessible areas. Ultimately, Geographic Information System (GIS) techniques can record data well beyond the purely geometric ones, to include statistics, intangible heritage, links with Land Registers, cultural aspects, archive resources, stratification of historic data, etc.

Archaeological surveys are required in sites with a series of superimposed layers of history (Fig. 4.3). In some types of conservation areas and archaeological parks they are mandatory but in general they can clarify the inter-relations between phases in complex fabric, both in plan (especially underground) and along the elevation (mostly relevant with medieval archaeology). The stratification method records the extent of each stage and maps them in a matrix in order to decipher their succession and whether they are destructive in reference to the previous ones.

As mentioned above, these surveys produce broadly two types of data, with the quantitative ones being more suitable when the state of the building can be described with precision and small changes may be critical for

4.3
Example of an archaeological survey from the Nybster broch in Caithness, Scotland (Courtesy AOC Archaeology)

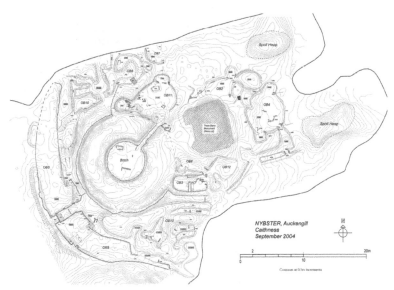

its stability. Together with the measured survey techniques, there are several more scientific methods, some developed in laboratory conditions, which are examined in the next section. Qualitative data (location and direction of cracks, buckling, stone degradation, etc.) can be recorded in each of the other types of surveys and there are several straightforward types of notation that help the professional to quickly monitor the problem. Table 4.2 is an example.

4.3 Data surveys

The various survey techniques are presented below, according to the quantity measured rather than the technology or type of datum these techniques are based upon. These quantities are broadly subdivided into global displacement, deformation of the fabric (and the often associated assessment of stress) and integrity of the cross-section or surface. It has to be remembered that because of the often imprecise or heavily deformed nature of historic construction, the tolerances are high and readings must be interpreted within the overall accuracy of the fabric or the context of (natural) long-term deformations. Beckmann and Bowles (2004) and Alvarez and Gonzalez (1994) present a more extensive discussion on the instrumentation available, its application procedure and its limits.

4.3.1 Displacement and distortions

The exact measurement of changes in length, deflection and out-of-plumb has two purposes: to establish current condition (and validate structural models) or to monitor ongoing deformation (Beckmann 2004). They have to refer to the original configuration of the building or to one that is well known. Choice of location is fundamental if the measurement is to represent the response of the building and requires some knowledge of the quantity already, such as its magnitude or direction.

Displacement can be assessed against an original datum if the building's initial configuration is known with confidence, which is often difficult. Most important is movement that is still developing or is expected to fluctuate according to certain conditions (such as traffic or temperature changes in a bridge). In the first case, the measurement helps to interpret

Table 4.2 Notation of key defects (after Carbonara 1990; Andrews 2003; Beckmann 2004)

PATHOLOGY		SURVEY	MATERIALS
Visible crack	Fissure	Out-of-plumb reference	Concrete
Crack at the outer face	Diffused cracks	Core samples	Brickwork
Through crack	Plaster detachment	Soil or foundation samples	Limestone
Slanted through crack	Area of degradation	Photographs	Sandstone
Crack formed by joining fissures	Humidity		Plaster
Bulging cracks			Mixed fabric
			Generic masonry

the deformation of the building, while the second serves to assess how critical the condition is (whether it has exceeded a notional serviceability limit). A range of such deformations, as directly measured or assessed analytically, and their effect on the stability of vaults was presented in Tables 3.5 and 3.6.

Placing a dial gauge, an Invar tape or an LVDT (Linearly Variable Displacement Transducer) requires first an assessment of whether the displacement is critical and what is its principal direction. Its magnitude has to be estimated in order to choose the right length of the gauge but above all it has to be mounted in complete independence from the deforming structure. The method clearly has laboratory origins and these difficulties may make it unfeasible.

The measurement of settlement is more meaningful and from single points the overall picture of the deformed structure can be pieced together. They can be either related to the undisturbed soil through a carefully fastened datum line or taken directly against adjacent (and less disturbed) members to evaluate the differential settlement that causes shear deformations. Direct reading of settlement can be combined with the reading of levels and an accuracy of 0.2 mm or even 0.05 mm is possible. Self-levelling instruments used to be very useful as they could adjust automatically, when compared to very accurate bubble levels. Out-of-plumb measurements are also required to assess eccentricities of a structure. Where direct readings with a string plumb line are not feasible due to difficulty of access or the number of locations, optical methods can be set up, where a bolt is fixed and its position is recorded to 1 mm accuracy by a theodolite (Beckmann 2004).

The precision of the instrument and inherent problems of the method should be well established by the manufacturer or the theory, and there are also strategies that allow errors to be mutually cancelled. Ultimately accuracy depends on the observer and how easy the instrument is reached and read. Equipment that translates changes in length or level into voltage (LDVT, electrolytic levels, strain gauges, etc.) can be monitored remotely and stored in hard drives. Such instruments can read to high precision but nowadays all such information can be collected and indirectly assessed by laser surveys. Depending on the accuracy required, it is important in that case to use many stations close to the defect and increase the density of the scanner. Such arrangements may increase the time to collect and analyse the data but the quantity and quality is such that it compensates the costs.

Dynamic actions are another critical area of monitoring. Such actions result from earthquake tremors in seismic zones and, more universally, vibrations from traffic or industrial uses. Similar to the static data, what are assessed are quantities that indicate a critical condition of the building, such as acceleration of parts of the structure against the (often) known ground acceleration. Otherwise, properties of the structure that allow validation and set-up of analytical models are also measured, for example, the translational mode periods and their variations, which are eventually used in the numerical evaluation of the eigenmodes of a structure.

These problems are typical in very slender towers (bell towers in Italy, such as Monza or Pavia, or beacons in the Danba valley, or in Szechuan, China), Byzantine churches in most of the East Mediterranean region or, even, more modern buildings such as the City Hall in Los Angeles. Accelerometers are compact lightweight units with an internally suspended mass that remains stationary while the body of the transducer follows the movements of the structure (Beckmann 2004). They are sensitive to higher frequencies and when they are part of a seismic strong-motion measurement system they can record within the range of 5–50 Hz which represents the limits for civic protection.

The key issue is the meaning of these measurements. In historic buildings with complex geometries and superposition of variable masses, often there is no sufficient continuity, therefore the natural frequencies or periods in one part might not represent the behaviour of the building. There are also some more fundamental problems: in slender structures (medieval towers or Gothic spires that are affected by traffic or bell-ringing) deducing the basic modes of oscillation and consequently the natural frequencies is straightforward. In more squat structures, however, that are braced from a variety of lateral structures often of similar proportions, careful planning of the measurement in each of the parts is required and then composed so that an approximation of the dynamic response is deduced. Often natural frequencies can be obtained by carefully applied vibrations that allow a direct correlation with the readings.

The issue of measurement techniques and their meaning is certainly as broad as the structural types and their problems and has been covered in this section only at a level of awareness. Finally, the last important aspect is the continuous monitoring of a defect. Accessibility of the instrument, its long-term stability, the repetition of the set-up, ease of data retrieval, etc. are the critical parameters in planning and choice. What characterises many of these systems is that they function at low voltage which means durability of the battery and small units. Collection of the data can also be done remotely, even through a mobile phone, and software allows controls to be set and alerts to be sent if extraordinary values are recorded.

4.3.2 Deformation and stress

This kind of information represents the effect of internal loads in the structure. What is measured is direct stress within the fabric or its strength through samples, and the manifestation of strain and failure through the assessment of cracks. Many of the previous practical issues apply here, especially how representative the readings are of the problem.

Strain under a specific load can be directly assessed in continuous solid materials such as metals and hardwoods by strain gauges or in any other materials with a highly precise demec (Demountable Mechanical strain) gauge. The first method measures the change in length of a small resistance fully glued on the material, through a Wheatstone bridge circuit expressed in micro-voltage (μV) and then translated in micro-strain (10^{-6}) and recorded through a data logger. The full contact of the gauge with the material is fundamental for the readings to be meaningful, which is addressed by the preparation of the surface and application of the gauge. The other method measures directly the change in the span between two fixed points by manually applying a relatively accurate dial gauge. The precision of the method depends on the ability of the surveyor to apply the gauge with a sufficient force and in a repetitive manner, which might be critical in small deformations. In either method, if the modulus of elasticity is known and fairly uniform, the strain can be transformed into stress.

In masonry structures stress and strength are often measured by a moderately destructive test using flat-jacks inserted in a carefully opened slot in the fabric. As can be seen in Figure 4.4, they are made of a hydraulic jack fixed on top of a flat metal plate which inflates a small metal spreader beam on top against the fabric. The principle is that the force the jack exerts on the masonry is equal to the stress required to bring the slot to its original configuration, or to the strength if the maximum force is recorded.

There are also other historic methods to assess directly the strength of a new structure by loading using sandbags or even the workforce spread in regular locations along the structure (as E. Torroja or E. Dieste did in their shells).

Laboratory tests are more destructive techniques that require samples taken from the fabric to be tested under load to failure by compression or shear. Cylindrical samples can be taken by core drill tests while sometimes undisturbed wallettes can be carefully removed and tested, especially for shear or bending.

4.4
Reading stress with a flat-jack (martinetto piatto) – courtesy CONTROLS S.R.L. – Italy

Finally, it is important to assess the real width of a crack and whether it is still developing. The first issue needs to be clarified before any variation can be detected. Apart from structures where crack width can be interpreted by a reliable theory, as in the case of RC structures, often it is not the actual width that is assessed but whether it is still widening or simply fluctuating with seasonal temperature or moisture variation. The real edges of the crack have to be identified first as they might have been concealed with later repairs. Then, the measuring device can be located correctly on these edges.

Various types of instruments can be used, from the most direct ones such as 'butterflies' and tell-tales that can break easily when the crack moves; glass plates normal to the crack that can move differentially and indicate magnitude and direction against a cross-line, which come in various systems (copper, aluminium, acrylics) and trade names such as Avongard (Beckmann 2004). The crack width can also be measured directly by demec gauges (where points are fixed on either edge), precision callipers or permanent, more accurate instrumentation using LVDT devices that read and record the movement along the direction of the transducer (Fig. 4.5). The latter require a data logger and careful set-up but are very useful for repeated measurements or inaccessible areas.

4.3.3 Integrity and condition
This last category does not refer to the geometric stability of the structure but mainly the condition of the fabric. What is explored rather than measured in the main traditional systems can be summarised as flaws and voids

4.5
LVDT monitoring of cracks in the church of San Julian de los Prados, Oviedo

in metals (from the casting process); voids within a masonry wall (in a ruin or degraded rubble core that has suffered severe loss of mortar or binder), the presence and conservation of reinforcement in RC structures; moisture content; insect attack in timber.

Cores taken for strength assessment can also reveal the presence of such defects. The method, however, is clearly destructive, the location and extent cannot be known until the samples are examined, and often many samples from inaccessible areas may be required.

A series of methods under the general name Non-Destructive Techniques or Evaluation (NDT or NDE) provide feasible alternatives and a range of cost-effective rates. The basic principle examines the way fabric reflects a wave emitted by an acoustic or electromagnetic source, with high velocities indicating a material of good consistency and fewer voids or cracks. Apart from the porosity, hidden features, defects, etc. the velocities can be translated into material properties (chemical, mechanical, physical) and eventually characterise the material, both in its present state and towards its conservation. The main ND methods will be briefly outlined (McCann and Forde 2001; IAEA 2002), discussing their applications and limitations regarding historic structures.

Ultrasonic testing is based on very short longitudinal ultrasonic pulses emitted by an electro-acoustical transducer applied on one face of the material under investigation. The pulse is transmitted through a liquid coupling material (such as grease) at the interface of the transducer and undergoes multiple reflections at the different material phases or features of the fabric. Longitudinal stress waves reach first the receiving transducer and are converted into an electrical signal, eventually allowing the pulse velocity to be measured. Through experience, empirical relationships have been established between the pulse velocity and some mechanical properties.

When damage or defects are present, the pulse velocity is reduced and, if mapped, an indication of the extent of the defects can be evaluated. Good compaction or solid material is shown as an increase in the pulse velocity. The technique is more suitable for metals and alloys, but it can be used in porous materials (concrete, stone) or timber, to various degrees of resolution.

Electromagnetic Testing is based on electric currents (principally eddy-current effects) or magnetic induction applied to the fabric. The electromagnetic response can reveal an internal defect according to its depth, with near-surface cracks and metal corrosion being more easily detected by Eddy-Current Testing (ECT). The latter is more effective for non-ferromagnetic materials, as otherwise the penetration is relatively shallow, in which case reinforcement can be detected under a thin cover.

Radiography provides information on the composition of the fabric: the X-rays or gamma rays emitted by the transducer lose intensity as they pass through the material and are absorbed or scattered. The amount of radiation emerging is recorded on a radiation-sensitive film and the loss is associated with density of the material and its thickness, which allows for material characterisa-tion. A key limitation of the method is the case of thick sections, which demand high-energy radiation, requiring powerful and heavy equipment emitting for long periods of time, a process with cost and Health & Safety implications.

In the same family of techniques, Ground-Penetrating Radar (GPR) is a useful and rapid geophysical method based on using radar pulses to map underground objects. It is a popular technique for archaeological investigations and can clearly detect voids in historic structures (Fig. 4.6). The method is based on the propagation of long-wavelength electromag-netic radiation through materials of different dielectric constants (IAEA 2002) and measures how much of this energy is reflected from any subsurface features or dielectric boundaries. The high-frequency radio waves are trans-mitted by an antenna and a receiving antenna records the signal reflected back. The greater the difference between dielectric constants (from two different adjacent materials), the greater the amount of energy reflected and the smaller residual energy propagates to the other material.

Limitations in the depth penetration and resolution depend on the medium's (soil or fabric) electrical conductivity and the radiation frequency. As the energy is more quickly dissipated into heat, high conductivity or wet media (moist ground, dampness in structures) decrease the signal strength and eventually the penetration depth. The same problem occurs also with higher frequencies but they have the benefit of higher resolution. Radar waves can easily penetrate dry sandy soils or building fabric but in wet soil conditions, clay, or good conductors such as metals, penetration is very superficial.

Acoustic emission methods are based on a natural property of any material to produce a sound or acoustic waves each time a mechanical disturbance or a crack occurs. The signal is then detected by piezoelectric

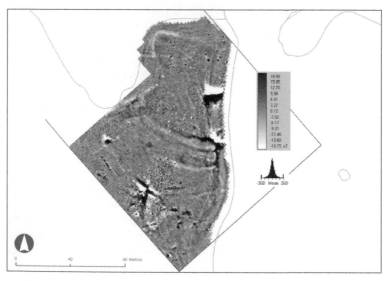

4.6
Results from a gradiometry survey of later prehistoric defences at the promontory fort at Cults Loch, south-west Scotland (survey by Tessa Poller, courtesy AOC Archaeology)

transducers that convert it into electric signals (IAEA 2002). The source of the signal (and therefore the issue or its depth) is determined by the time the wave takes to travel to the receiving transducer. Sometimes the mechanical disturbance can be stimulated by a carefully controlled loading.

Finally, infrared thermography is based on the naturally occurring phenomenon that all objects above 0 °C emit infrared radiation. This radiation is expressed as temperature distribution at the external surface of a body and is detected using infrared cameras. The composition of the fabric or soil certainly determines how heat is stored and emitted, and a careful analysis combined with some knowledge of the materials and consideration of the environmental conditions permits the characterisation of the fabric. Variation of temperature with time can also reveal the presence of features such as voids or different materials to be detected, if the rate at which they dissipate heat is known.

Chapter 5

Strengthening and repair techniques

It is important to discuss the techniques that can improve or simply restore the strength of the main historic materials, with reference to their mechanical properties or their behaviour in single elements, without directly dealing with the overall structural scheme or its condition. The technology and principles of these repairs are discussed alongside their structural design potential or limitation, which will be further reviewed in the applications of the next chapter.

5.1 Materials

5.1.1 Masonry

In stonework, the most effective and passive decision is the choice of stone. As seen in the previous sections, the more urgent problems in interventions have to do with physical compatibility, the effect of colour choice or tectonics rather than strength. This is the usual case in replacement of decayed stone, as often happens in Georgian buildings in Edinburgh, or the addition of new ones (in extensions, completions or repristinations). While it is possible to reproduce the colour, size and texture of original bricks, stone is usually replaced primarily in terms of physical and mechanical compatibility (see properties in Table 3.1), and then colour is matched. If new stone is carefully chosen to have similar properties, such as durability, strength, expansion, etc., to the original, it is always difficult to get a very similar colour, even if it comes from the same quarry.

The procurement of stone in this case is of fundamental importance. In Scotland, for example, in the nineteenth century each town had its own sandstone quarry producing stone of various but also variable qualities. Only six quarries exist today but this is not the result only of industrial changes but also stronger demands in quality, durability, etc. Ease of extraction and minimisation of waste are other parameters that affect the viability of each quarry.

This also affects the secondary treatment of stone by the masons. The availability of high-strength tools and mechanised procedures allow higher control on the dimensions and finishes, a particular feature in historic masonry which is distinct from the more standardised production of stone for modern cladding systems. This also affects the durability of the stonework in service conditions, as sedimentary stone (such as sandstone), in particular, must be placed in a specific orientation (Fig. 3.20).

Dressed and rubble stonework (the load-bearing part in a composite wall) have quite different damage process and repair methods. Careful substitution is possible in the former, to improve mainly durability and aesthetics, while the latter often requires partial reconstructions that result in strength improvement. Similar issues apply also to brickwork, although in this case bricks can be more easily fabricated to match dimensions and texture of the original.

Repointing and reconstructions are repairs that can alter the aspect or even originality of the fabric. The theoretical issues discussed in Section 1.2 (Reintegration of the structural system) and 1.3 (Materiality) apply precisely here and ultimately the focus of this book is towards a designer's informed choice on the extent and selection of materials. The ability to distinguish the new fabric from the original must not be at the expense of the aesthetic unity of the work (Fig. 5.1) but also the historic values must be readable (Lupp *et al.* 1998).

Reconstructions can be carried out by first removing the damaged fabric and cleaning every loose joint. If a tight fit is required (ashlar stonework, high-quality terracotta finish) jacking of the overhanging courses may be required to allow insertion of the new blocks. Reconstruction may also be a solution when cracks that do not move any more have to be repaired. More than the strength, this repair can improve the durability of the fabric. More specific solutions according to the loads that have caused the failure or the location of the wall will be given in the next section on load-bearing elements.

5.1
Matching of new
sandstone in a
Georgian building
in Barony Street,
Edinburgh

Repointing, apart from protecting the joint (usually more vulnerable than the unit) also restores the full, load-bearing cross-section of the wall, spreading the loads more evenly (a repair often completed by grouting). The joint must be thoroughly cleaned by chisel or a grinder, then washed and the mortar or grout must be deeply applied to fill the void (Fig. 5.2). The choice of mortar is based on compatibility with the behaviour of the masonry and often lime-based mortar must be specified for its well-known durability and flexibility. Such properties permit the accommodation of small load changes and the avoidance of cracks, but equally important allow the masonry to 'breathe' any moisture outwards. As we have seen, the use of harder, cement mortars can cause the erosion of the unit, because of a higher impermeability compared to stone or brick.

Lime mortar injections are recommended where significant binder or volume has been lost inside the load-bearing section. Closely spaced tubes are fixed on the accessible face of a wall (Fig. 5.3) and the grout is injected at medium pressure until a prescribed pressure is measured or grout comes out from an adjacent tube. As mentioned earlier, rheology of the grout or the setting of non-hydraulic lime within very confined conditions is a major problem, further complicated by long setting times. This has major implications in very wet climates or in precarious structures, especially in unstable ruins because, since strength is not achieved rapidly, costly temporary works are required. Current research aims to increase fluidity by using small-scale additives or to accelerate carbonation naturally without disturbing the cavities by applying high pressures.

5.1.2 Timber

The restoration of timber focuses on two key aspects: cancel the sources of decay and substitute damaged sections. As with masonry, aspects linked with particular loads and functions will be dealt with in the section on beams

5.2
Repointing in rubble
stonework

5.3
Grout injections
to strengthen the
masonry vaults of
St Francis in Assisi
(courtesy Dr Alberto
Viskovic and Studio
Croci)

and roofs. Drying a very wet timber or killing any biological attack has to be done very carefully but it is complicated to reproduce laboratory conditions on site or remove the damaged timber to the lab. Rapid shrinkage during drying or killing of organisms (algae, fungi or insects) can cause distortions to the fibre and even collapse of the cells, resulting in localised damage. The rate of these actions therefore has to be carefully controlled.

Especially in the case of dry rot, heavily affected timber might have to be removed and replaced if it cannot be dried out (Beckmann 2004). Parts of a beam or post might have to be replaced for many other reasons, such as past mechanical damage (from bolts or notches), splits, knots, etc. A light temporary prop might be sufficient, but where an entire member has to be replaced the key problems are the operation to remove it (while ensuring support of the loads) and the insertion of the new member (Beckmann 2004).

As in stonework, timber will have to be chosen for its compatibility characteristics rather than a mere similarity in terms of species. The procurement of new timber, then, becomes less critical as far as authenticity is concerned. Visual grading (to define the size, type and number of strength-reducing characteristics allowed in each grade) may be more straightforward and relevant for the definition of these properties than machine grading (which determines exactly the strength and stiffness). Figure 5.4 shows

5.4
Scarf joint for
replacement in timber
(CIRIA 1994)

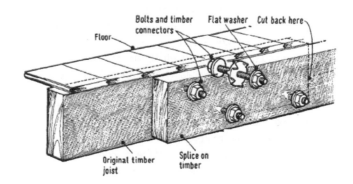

Bolts and timber connectors   Flat washer   Cut back here

Floor

Original timber joist   Splice on timber

how to attach a new piece where mainly compressive forces are expected and other arrangements exist for tensile, bending or shear forces. The problem essentially is not too different from the original historic timber joints that had to provide strength and continuity for the same forces.

The availability of trained craftsmen (joiners, carpenters) is another important factor in the addition of the new pieces or the re-creation of ancient structural forms. This was the case in the full reconstruction of the hammerbeam roofs at the Great Hall in Stirling Castle, Scotland (see Section 6.4). In this case, as in other important restoration projects, timber grown in special forests is used. Grading should also take place for the selection of the compatible and stronger pieces and, as in modern timber structures, they should be accordingly placed in the frame or the roof. Finally, it is important to remember that any new timber structure will eventually shrink, therefore the new timber should be seasoned or should have a similar moisture content to the existing wood.

Strength can be improved using distinct materials, such as steel plates that can be fixed externally or in the core of the beam through a slot (flitch beam). Resin-based reinforcement is essentially a variation of the flitch beam where metal is substituted by sandwiched plates containing FRP stripes or bars are inserted in the slot and then fixed with epoxy resin.

So far as fire resistance of timber is concerned, fire safety engineering practices should be first planned (ensure fire does not reach vulnerable elements, provide barriers to the spread of fire within a member, compartmentalise the building to reduce fire load). The naturally good resistance of timber (the ability of the outer charred layers to protect the non-ignited core) can be further enhanced with intumescent paints that protect the member from direct effects. It is then essential to protect also the components with lower fire resistance, such as metal joints, or even to shield the timber members from their failure.

5.1.3 Iron and steel

The main goal of any intervention here is to ensure the robustness of the building as most historic applications of steel and iron have purely structural function and, apart from bridges, the steelwork is covered or has no aesthetic value. As will be seen later in the discussion of framed buildings, this robustness is achieved by strategies to conserve or improve the lateral stability and the integrity of connections.

Regarding the material itself, there are sources of problems due to mechanical reasons as well as material pathology. The repairs must improve any low structural capacity that may have resulted from early inadequate production processes, defects, increased loads, age, etc. or in some cases the source of the problem will have to be arrested (corrosion) or kept away (fire). Focusing on the material problems, corrosion can affect steelwork when water cannot run off the surface or connections. In exposed members it occurs when protection has worn off and incorrect detailing does not prevent water from lodging for long periods or toxic dirt from accumulating. In members encased within masonry or concrete, water can infiltrate from cracks or joints but cannot dissipate equally easily. Often the problem is hidden but can further accelerate once corroded steel expands and puts pressure on the encasement.

The repairs, first, must guarantee that such processes are arrested. Maintenance of ancillary areas can prevent water from lodging and simple inclusions of flashings around edges may improve water flow. Then the remaining section must be protected. Although rust may give an impression of great section loss, it has to be remembered that usually 10 mm of rust can result from the corrosion of just 1.5 mm of steel, which might not be catastrophic (Beckmann 2004). Any loose material should be brushed or sand-blasted, and once stable and clean metal is reached it has to be coated with waterproofing paint or given galvanic protection. Encasement should also be replaced with sound new terracotta or concrete cladding (as in the De la Warr Pavilion – Fig. 3.23). Radical measures, such as dismantling and reassembling can cause more serious difficulties as the geometry of the frame might have changed and reforming the connections in the original manner (rivets) might be impossible. Cathodic protection is another strategy to arrest and reverse corrosion but is more often applied to steel bars in concrete.

Mechanical repairs attempt to reinstate lost, damaged or weakened material or stiffen and enhance restraint of a section. It is very important to understand how they may overall change the loads regime (whether the original, damaged or altered) as often an increase in the section appears to be the most straightforward solution. However, a more effective remedy can be the relief of stress or re-direction of loads to other areas through changes in stiffness or even addition of new elements, preferably independent from the original frame. In some cases, the built-in stresses from the history of the building, which can result from unpredicted deformations, elongations and shrinkages, must be considered, and then relief of the load might simply bring the member back to normal levels of stress, within its material strength.

Strength can be improved by increasing the stiffness of the section by adding plates according to the applied load, or by repairing cracks and bulges. Cast-iron and wrought-iron members cannot be welded so any additional plates will need to be bolted, choosing carefully the type of bolts to increase shear strength and minimise the number of holes, such as high-strength friction-grip bolts (Beckmann 2004). In cast-iron members that are expected to receive mainly compression (tubular columns) strength can be improved by a collar that restrains the column. Where brittle cracks have appeared in a beam, they can be stitched with a pre-drilled plate (Fig. 5.5). In wrought-iron or mild steel beams, C-sections can be added symmetrically to a column or to the tensile flange to increase stiffness. Alternatively, and similarly to timber or concrete, FRP can be added in the same areas to reinforce the member with only minimum visual intrusion or drilling of holes.

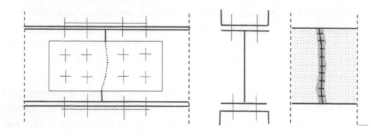

5.5
Cast-iron stitching
and addition of plates
(after Beckmann
2004)

In some cases, the structural design may show that full replacement with a new high-strength steel member could be the most feasible solution, reducing the number of repairs and relieving loads from other weak areas. The delicate process of removing and replacing must be planned in such a way that no additional loads are transferred temporarily to the adjacent (and probably equally weak) members. A key problem is how to match the sizes, for constructional and aesthetical reasons. Standard sizes have changed often in the past, while some countries, such as the UK, have changed to the metric system.

The industry could produce specific cross-sections for prestigious projects that can afford the cost, mainly for cast iron (a material still in use today) or mild steel. There is however a serious lack of foundries, especially for wrought iron, which in the case of the UK is all imported, in contrast to the past where pre-manufactured decorative units would be exported all over the world. The reproduction by welding or bolting plates can give results that are satisfactory structurally but not visually and artistically. In many cases a compromise must be made to what is available in the market, once the effect of steelwork on the aesthetics of the building has been established and understood, as was the case of the refurbishment of the curtain wall in Boots D10 (see Section 6.7).

As with all historic materials, compatibility in terms of thermal movement, strength and stiffness must be considered, beyond a mere similarity between the original and new material. If the budget of the project allows, a metallographic analysis of the original fabric can reveal the composition of the minerals that control its durability or the carbon content. This will allow the right alloy to be chosen and avoid problems such as bi-metallic corrosion or the redistribution of loads in unknown proportions.

Finally, protection of steel members in case of fire is often part of the more desirable design strategies to improve the load-bearing capacity of a historic building or update it to current safety levels (fire, increased capacity, accessibility etc.), as was the case of J. Stirling's Engineering Faculty in the University of Leicester (see Section 6.6). The most direct solution is to improve the fire proofing of the member with intumescent paints (nowadays in white colour), where the aesthetics and character of the building allows it. In some cases, tubes can be filled with water or even concrete, where they cannot be encased, a technique that is rather aggressive and irreversible while the important compacting cannot be fully guaranteed. Other techniques, such as the provision of barriers or the allowance for movement after a sustained fire will be discussed in the frames section below.

5.1.4 Concrete

Repairs should address the three main aspects mentioned in Chapter 3: strength, surface finish and construction process. In structural terms any intervention must ensure the robustness of the building and lateral stability and apart from the design strategies for the entire building, a repair can focus on the strength of the joints, aiming for a monolithic effect or ductility in cases like seismic performance. Often earlier design codes may have not considered certain loadings (e.g. accidental damage, increased imposed loads), therefore vertical continuity in joints and robustness must be provided. Loss of durability in concrete has various causes and manifestations as was seen in Chapter 3 (quality and compaction of concrete, aggressive chemical attack, corrosion of reinforcement).

In both this and the previous area, the repairs would often concentrate on the protective reinforcement cover or the external face of the member. Other strengthening techniques tackle the loss of reinforcement strength and the presence of cracks. Starting with defects and cracks, the less aggressive techniques are targeted repairs using hand-applied fluid grouts (epoxy resin, polyester, cementitious) pumped under pressure to fill voids and cracks or re-bond the material. The most aggressive and non-reversible technique is jacketing by spraying concrete over a reinforcement mesh, either as a dry mix (Gunite) or wet (shotcrete). Any of these techniques, however, have a limited life which can be extended by establishing a programme of maintenance.

Some of these techniques are also useful for extensive repairs where there is serious discontinuity in the fabric. There are techniques to repair defective concrete by replacing with polymer-modified mortar or concrete, resin mortars or thicker layers of sprayed concrete. If the damage to the reinforcement bars is extensive, new bars can be inserted or new external steel plates can be used (Fig. 5.6). FRP can also provide a less intrusive and reversible solution through fibre composite plate bonding or carbon fibre reinforced polymer composite strips can be applied for flexural and shear strengthening.

A more specific range of techniques attempts to arrest and reverse the effect of corrosion on the re-bars (see Fig. 3.14) by electrochemical protection that, essentially, makes steel a cathode. This can be achieved by fixing an external anode on the surface or embedding the anode in the concrete (Fig. 5.7). A DC power supply (the transformer rectifier) is essential to pass current between the anode and the reinforcement (the cathode). The main variations of the process are cathodic protection, electrochemical chloride migration or 'desalination', and re-alkalisation.

Cathodic protection, the best-known process, uses permanent anodes, power supply and control systems (Fig. 5.7). The anodic reaction

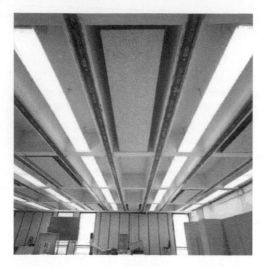

5.6
External repairs of
RC beam and roofs
(courtesy Sika Corp)

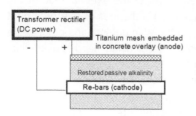
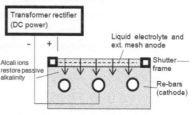

5.7
The mechanisms
of electrochemical
protection of steel
reinforcement in
concrete: cathodic
protection and
re-alkalisation

(the aggressive part) is isolated to a corrosion-resistant anode, while the cathodic reaction occurs at the surface of the reinforcement. The process essentially rebuilds the passive alkaline layer of the steel bars and repels any chloride ions. Re-alkalisation, on the other hand, consists of a temporary application (every six months) of liquid electrolyte with a mesh anode (contained within a shutter frame) that lets alkali ions through concrete, producing hydroxyl ions that eventually restore the passive alkalinity and protection of the reinforcement.

Aside from the material itself, larger-scale strategies to improve the strength or robustness of the structure may be required, such as cancelling adverse bending by post-tensioning systems. Such solutions managed to reverse the alarming deflections in Fallingwater and they will be reviewed in Section 6.5.

Regarding the repair from fire damage, the thermal mass of concrete around the steel bars protects them effectively against excessive temperatures. The provision of the right type of cover according to the fire exposure and use of the building are the essential types of protection. Where concrete has been damaged by the deformation of steel, it will need to be replaced using the techniques mentioned before and to the right type of cover.

Finally, an interesting aspect of concrete is its historic use as a repair material, even for masonry (Calderini 2008). Before the war, RC was considered in Italy as the ideal conservation material, especially when conservation was seen as a purely technical problem (Camillo Boito, Gustavo Giovannoni): it avoided falsification of the original masonry fabric and confusion, provided optimum structural strength and stiffness, sizes could be exactly calculated, etc. It was used in the restoration of the bell tower of the Basilica of San Antonino in Piacenza and the dome of San Gaudenzio, in Novara.

5.2 Structural elements

The repair strategies and design possibilities for entire structural systems within a building (roofs, floors, wall assemblies, etc.) will be discussed in this section. The global strategies for the entire building are presented mainly for frames and walls, including fire engineering approaches, and some more specific methods for vaults and arches are presented at the end. In this section, we will consider mainly repairs that maintain as much as possible of the original configuration and structural scheme of the element. Major alterations with a richer design potential are discussed in case studies in the next chapter.

5.2.1 Roofs and trusses

Historic roofs with interesting structural issues are mainly constructed from timber. Large spans are often made of steel and iron and are found in well-preserved railway stations, and they will be dealt with only briefly here.

Most timber roofs do not follow the truss scheme but are a combination of rafters with posts and ties. Large-span roofs, such as the ones in cathedrals, use good quality hardwood, which if not attacked by insects, humidity or fire remain in almost pristine condition. Like any frame construction, the overall strength and stability depends greatly on the connections and this is another area where the repairs may be focused.

Restoring the strength of individual elements using the techniques discussed in the previous section (repair of compressive joints, scarfing joints, replacement of rotten timber, reinforcement, etc.) spreads the distribution of the loads more evenly among all the members. The insertion of new metal shoes at tensile joints, hardwood pins or stainless nails also restores the function of the connection.

Such techniques aim to restore the original scheme and the 'regola dell'arte' to the standard loads affecting the roof. However, more severe conditions need to be considered, such as excessive loads (e.g. from additional insulation), accidental loads or snow, as continuous melting can soften the rafters. Making the roof impermeable reduces and prevents the direct effect but extra redundancies may have to be provided. As with modern timber, more moment-resisting joints or improvement of robustness can ensure that loads will be shared elsewhere in the structure if released by a sudden failure, and in this way disproportionate collapse is avoided.

Other interesting areas of problems result from the need to move loads away from the underlying fabric as much possible. In the case of large stone vaults, the roof is often supported halfway on the vertices of the vault. When the two structures are incompatible (differential deformations, excessive vertical loads and thrusts), they might need to be decoupled or the structure of the roof changed to reduce the thrust component (usually the most critical one). The addition of new members might be needed to share the loads or change the load distribution, while bearings can be mounted on the roof edge to release some of the thrust or thermal movement.

Following structural analysis under a wider variety of action combinations, the location of such members can be identified, ensuring that they would not develop adverse stresses. New steel ties can be added alongside existing under-performing ones, or in other positions that provide an effective bracing effect, as was the case in St Mary's Church in Sandwich restored by P. Beckmann (2004). The interesting solution of a new lightweight timber roof over the domes of the medieval church of San Agustín in Valencia, Spain incorporates a system of cables in the vertical posts and horizontal ties that allow the loads in each frame to redistribute when the rafters extend due to thermal movement.

Complete reconstructions of historic roofs also offer an opportunity to reflect on the efficiency of the original scheme and explore how far its rules can be obeyed and combined with effective modern solutions. The case of an extensive reproduction of the original geometry with new timber materials (Glulam) in Palazzo Carignano (see Section 6.4) is subtly but essentially different from the almost faithful reconstruction of the hammer-beam roof of the restored Great Hall in Stirling Castle using locally grown, Scottish oak.

When dealing with large-span roofs, usually the quality of the original construction is quite exceptional and, apart from maintenance of the material against rust and corrosion, their overall stability is quite sound. Large sheds like those found on railways often also show a geometric stability that allows them to be re-clad with modern efficient materials, as was the case of Paddington Station in London (Connell 1993): the roof was reglazed in 1992 with twin-walled polycarbonate sheet fitted in aluminium glazing bars and re-clad with British Steel's Plastistol colour-coated steel sheets. One of the most impressive examples is Dresden Hauptbahnhof refurbished in 2006 after the severe damage from the city's bombardment in 1945. The steel arches of the 1895 concourse were still in their place despite the destruction of the cladding and secondary structure, which allowed Foster & Partners to use them as support for the Teflon-coated glass-fibre translucent roof that protects passengers and controls the daylight in the station.

In smaller, domestic-scale buildings, structural problems in roofs are caused by the spread of the eaves. Bracing can be provided by ties along the edges, combined with improvement of the overall stability of the building. A traditional solution is timber bracing embedded in the masonry at the level of window lintels, while the thrusts from the eaves are collected and spread uniformly over the roof by a continuous timber plate. Sometimes such timber 'chains' can be found rotten from moisture within the wall and their effect can be restored by inserting new, treated and carefully damp-proofed timbers, metal open channels or even fibre reinforcement pasted in timber strips, which is also less visually intrusive. As mentioned, it is important that any solution is combined with improvements in the horizontal stiffness of the building, such as tying the corners or strengthening the diaphragm action of the floors.

5.2.2 Beams and portal frames

As with modern structures, the stability of a frame building in any of the historic materials can be improved either by overall strategies or specific actions according to the type of loads affecting an element. A passive yet efficient approach is to relieve the fabric from as many of the current or added loads as possible, by adding new members or modifying the stiffness of the existing ones or the entire building. The addition of stair or lift shafts is usually a modern requirement in restored buildings for public use (museum in Palazzo Altemps in Rome) and even retrofitting existing shafts (Pirelli Tower in Milan) can contribute to the exact determination of the stiffness of a building.

Stability can be further improved by combining these stiff cores with increased in-plane stiffness of the frames. This is achieved by carefully introducing diagonal bracing or strengthening the joints (to ensure moment-resistance). Often, however, a new steel or precast frame is inserted (Tate Modern, London), especially in masonry buildings, to guarantee strength to the new loads and, of course, more useful space. Balancing the ethical questions or the delicate process of inserting such structures (Section 6.9), there is a benefit in that such a new frame can be precisely sized and analysed. Such methods can be combined with interventions to floors to restore their vertical and horizontal (diaphragm) load-bearing capacity, as will be seen in the next section.

Focusing now on the contribution of individual members, the specific characteristic of strengthening techniques in timber or RC beams is that they often involve a different material. The design choice is then a matter of how far they will be visible or technically intrusive, while considering the theoretical issues on the authenticity of the fabric and original performance (see Section 1.5).

Depending on the accessibility and historic and artistic value of a timber beam, the strengthening can be completely contained within the section (Fig. 5.8) or extra stiffness can be provided by extending the section (Fig. 5.9). Steel bars and glass or carbon fibres can be inserted by drilling and then anchored at the ends and fastened with epoxy resins. Plywood bands with embedded fibres can be inserted in a slot along the axis of the beam or

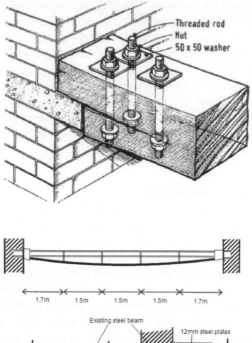

5.8
Reinforcement of a
timber beam (CIRIA
1994)

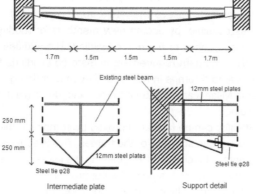

5.9
External
reinforcement of an
existing steel beam
as a trussed girder
(from the restoration
of Palazzo Altemps,
after Croci and
Scoppola 1993)

even on the external faces. This is a modern, slightly less intrusive version of a flitch beam with a metal plate, which can be further exploited if a new timber member is fastened next to the existing beam to double the original width.

A visually intrusive but fully reversible solution is to transform the beam in the compression element by adding an offset steel truss at the tensile area (bottom) that increases the lever arm of the section and therefore the strength of the beam (Fig. 5.9). This repair, as well as those solutions that locate fibres and rods at the tensile area, are clearly designed to withstand bending forces. A denser and more intrusive arrangement of the fibres in a diagonal mesh might be required in the case of higher shear loads, especially at the supports, areas of sharp geometric changes or where heavy point loads are received.

Some of these techniques can also be applied in RC beams in order to minimise intrusion. In most cast RC frames the problems are more of a serviceability nature (surface cracks, deflections, spalling, discoloration) rather than stability (more of a problem in precast systems). In some cases, the repair solutions presented in the previous section can be masked by a screed or even Gunite jacketing. Apart from visual intrusion and irreversibility, such aggressive finishes may add a significant thickness that changes the proportions of the frame or the openings, causing alterations or even load transfer to window frames. And they certainly must be avoided where the exposed concrete (texture or marks of the formwork boards) are an integral part of the original design (see the work of Louis Kahn).

The conservation works at the Park Hill Estate in Sheffield (built 1957–61) summarise many of the problems and repairs that arise in exposed RC frames (Beard 2001). No problems of structural movement, ASR or sulphates were detected but mainly spalling due to corrosion and carbonation of the mild steel reinforcement. Refurbishment of the plain tectonics of the concrete frame and the envelope (the graduated colour and texture of brickwork) were the key technical aspects in this project. The solutions sought were to attempt to arrest corrosion (by applying coats of corrosion inhibitor) and to restore the visual effects of exposed concrete by a cosmetic mortar. After the damaged areas were cleaned and repaired, and several options were examined, the best solution was to apply a water-based coating (aquaprimer) for the corrosion. Then a colour coating with elastomeric finish was applied in order to replicate the characteristic brutalist board texture of the exposed concrete.

Steel frames have fewer tectonic problems compared to timber and concrete frames but more crucial stability and robustness issues. As was seen in the previous section, improvement of connections is very important for the transmission of forces and function as a frame. The addition of plates in wrought or cast iron, the conversion in spliced beams or techniques to stitch fractured cast-iron beams might be acceptable in areas where the elements are not visible or are clad.

Repair strategies that link the strength of the beams (bending) or columns (buckling) with other parts of the building, however, have more important design implications in an intervention. External strengthening by the addition of a tie rod underneath the tensile flange creating trussed girders (Fig. 5.9) can alter significantly the perception of the space and may be more suitable for large spans or well-lit interiors. Enhancement of the composite action between a steel beam and a concrete or masonry decking made of jack-arches is a more compatible solution, where this configuration exists. The strengthening focuses on the stiffening of the solid deck and the improvement of the shear connection between the elements (Beckmann 2004) and is a discrete and effective solution in the conversion of industrial buildings, such as textile mills, into flats or offices, which often apply higher imposed loads.

The skeleton itself might be subject to disproportionate collapse especially in the case of cast-iron 'box' connections that are brittle and do not permit an even redistribution of the loads. Redundancies can be provided by adding extra members (such as new columns) which will then require continuity all through the structure and their own foundations. Alternatively, if the connection cannot be strengthened or even pre-stressed internally with a ring, allowing 'bridging' to occur can be investigated, i.e. the ability of the deck to span as a catenary following failure (Beckmann 2004). The resulting large deflections are the penalty for the maintenance of the overall stability of the building.

5.2.3 Slabs and decks

The key issues of floors in relation to the overall stability of the building are their load-bearing performance, the connections with the surrounding walls or beams and the diaphragm function. Most of the repair and strengthening methods probably will not affect the architectural character of a building considered in the design of an intervention, except in cases where the ceiling is visible or a specific thickness must not be exceeded.

Vertical load-bearing function can be improved with the techniques mentioned in the previous section for beams. In timber floors (Fig. 5.10) sharing of the load between as many of the secondary joists as possible helps reduce the deformation. Stiffening the actual deck with plywood sheets or metal straps over the joists could contribute to spreading the loads more evenly over the floor. Localised repairs might be required underneath the areas where high concentrated loads are expected by adding trimmer beams between the joists. Herring-bone bracing (braiding) is often found in timber floors in houses at one-third points along the span (CIRIA 1994) and a solution along these lines can be introduced in floors without sufficient in-plane stiffness. Timber enables some even more complex forms, such as the almost reciprocal arrangement of short beams to create a long-spanning 10 × 20 m floor for the Prospect Room in Wollaton Hall, Nottingham (built between 1580 and 88). The grid was deemed dangerous so in 1956 it was hung from three steel trusses, but a solution that integrates tension cables within the timbers was proposed recently to restore the direct load-bearing function of the timbers (Jolly and Brown 1999).

5.10
Repair of a timber floor and its support on a masonry load-bearing wall (Lourenço 2005)

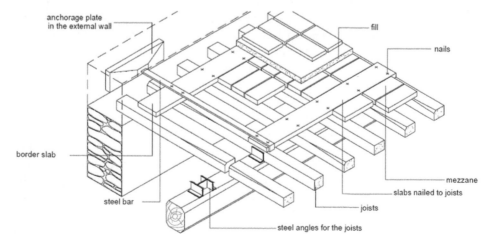

anchorage plate in the external wall

fill

nails

border slab

mezzane

slabs nailed to joists

steel bar

joists

steel angles for the joists

This is the serious problem that affects the performance of concrete or composite floors in RC or steel frames. Tying the building to the surrounding load-bearing walls or frame with tensile rods or even post-tensioning (in the case of the large cantilevering deformations in Fallingwater) improves the diaphragm action of the deck as well as its anchorage to the building. The shear connectors in historic jack-arches in a steel frame can provide a similar effect. Finally, the wall-to-floor connections can be improved by directly strengthening the support area using steel straps fixed between the wall and the joist or embedding the fibre reinforcement of the beam. Alternatively the support area can be strengthened by a continuous strap in steel, fixed along the perimeter of the floor or within a rebate.

5.2.4 Walls and plates

In this section, mainly masonry and RC wall assemblies will be discussed. Strategies to improve their load-bearing function under vertical and lateral loads will be presented. A key issue in their strengthening is whether stiffness is restored or a mixture of flexibility and ductility is introduced that allows the wall to redistribute differential deformations or large loads such as seismic activity. Finally, the appearance of a wall, especially in RC, is a technical problem related to the choice of concrete and its performance and durability.

Repair of a cracked wall usually needs to be studied within the context of the entire building. Once the cause of the damage has been relieved, the techniques described in 5.1.1 can be used. Stitching of the fabric should be avoided as it may simply migrate the problem somewhere else, and reconstructions of the damaged area should achieve a strong bond with the surrounding sound masonry; exhibiting physical and mechanical compatibility but also visual distinction.

Regarding strength, shotcrete solutions such as Gunite are aggressive to the historic and structural character of the fabric while the huge stiffness added can alter unpredictably the distribution of loads in the building. Injections, preferably in superfluid lime, provide repair of small cracks and voids with the necessary flexibility that allows walls to accommodate small movement without large-scale damage.

In-plane stiffness and tensile strength against lateral loads (from external forces such as wind or unbalanced forces from the floors or adjacent buildings) can be improved by stronger bonding and repointing. Reinforcement with FRP bands and textiles can alter significantly the performance of masonry, especially in a historic building, and should be used in extreme situations. They have to be applied at the area that is expected to bulge and unless there are aesthetic features (frescoes or ashlar finish) they can be covered with a light screed and plaster.

The outward stability of a wall, especially under lateral loads or after a small bulge, may require an overall strategy of lateral restraint. Stability can be improved by anchoring the wall to the corners with tie-rods

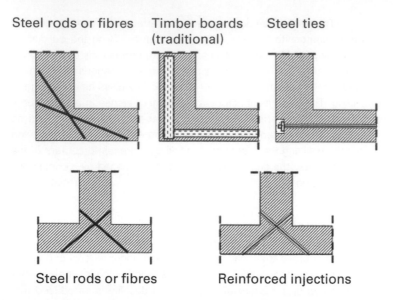

Steel rods or fibres | Timber boards (traditional) | Steel ties

Steel rods or fibres | Reinforced injections

5.11
Tying the corners
and crossings of a
masonry building

embedded in the wall (Fig. 5.11) or the floor joists. Anchorage becomes an even more critical technique in retained façades that have lost their intermediate bracing due to the destruction of the internal structure. The less technically demanding method of pier buttressing might also be a designer's choice that stems from a more traditional emphasis in a project.

Finally, walls that have seriously bulged cannot be returned to their original configuration but have to be rebuilt, also taking into consideration how the attached structures have moved following the deformation. In smaller deformations, tie rods attached on the floors can further restrain the outward moving wall.

The overall stability of the building against lateral loads, earthquake, etc. depends substantially on the bracing of the walls and how well they are tied. This strength is further enhanced by the diaphragm action from floors and ground, so vertical and horizontal ties must be provided or any existing strategy (traditionally in seismic zones through perimetral timber chains) should be refurbished.

Some other passive strategies are the removal of load from a delicate wall or one that cannot be strengthened intrusively because of important decorative features in its surfaces. Any excessive load from some returning walls or the roof must be isolated by careful detachment and insertion of dampers, but retaining some inherent structural stability is of fundamental importance. On the other hand, the addition of vertical load in some cases helps to stabilise a wall, as in the case of high thrusts in cathedrals which are rectified by the pinnacles or the repair of spire towers, such as in the Holy Trinity Church in Coventry (Beckmann 2004).

Some of these issues also apply to RC walls, and repairs of cracks or the cover can often be hidden by a screed or grout. Colour and texture match are very important in exposed walls, however, and these make up an integral part of the design in a conservation project, as was seen in Park Hill Estate. The repairs on the spalled cover of the vertically board-marked RC walls in Beethovenhalle in Stuttgart (Beckmann 2004) followed carefully the lines of the boards, and the sand for the new concrete was chosen from the same source as the original. Matching the colour and texture in exposed concrete needs specialist advice to identify the exact ingredients of the original mix and attempt to reproduce this by balancing the proportions of sand, cement, aggregates and water. The case of Einsteinturm in Potsdam (Hoh-Slodczyck 1999), another iconic modernist RC building, bypassed the problem as the concrete of the tower was rendered to give a uniform appearance.

A more serious combination of aesthetic and structural problems became apparent to the engineers who restored the characteristic concrete bands of the envelope of the Guggenheim Museum in New York. The spiral structure of the ramps was specified to be without movement joints by F. L. Wright and is made of sprayed Gunite walls strengthened by steel tee-sections. The stability of these walls was compromised by voids at their intersection with the web walls (Meade and Hudson 2010). The repairs included CFRP in the walls which were fixed to the rest of the structure with steel brackets and viscous dampers. The important exterior concrete finish was restored to match the existing, and further unified with a final coating.

5.2.5 Improving the overall stability and stiffness of the building

Returning to the theme of overall stability, the distinct systems discussed previously are integrated into design strategies that can provide, among other things, global lateral constraint or protection against earthquakes, and fire engineering protection, which can alter significantly the existing scheme of a historic building. A summary of such measures and their possible effect on a Georgian terraced house (typical in Edinburgh), for example, is illustrated in Fig. 5.12.

Significant restoration projects often require temporary support and containment of the forces released when the structural continuity is disturbed during repair works. They are visually intrusive but this is acceptable as they are temporary. Ties have been used temporarily to brace the north end of the Basilica of Maxentius in Rome against traffic vibrations that have increased after archaeological excavations at the outer base (Fig. 3.28). A system of tubes is anchored on demountable pads at the springings of the arches and vaults and in stiffened frames mounted in the openings (Carbonara 2005). The sophisticated system to pull back the Tower of Pisa to a stable inclination using cables was monitored constantly but required heavy plant on site to apply the forces (Croci 2006).

Temporary formwork for the strengthening of domes and vaults requires a quite dense scaffolding, which represents a substantial part of the cost of a project. The temporary scaffolding used during the strengthening

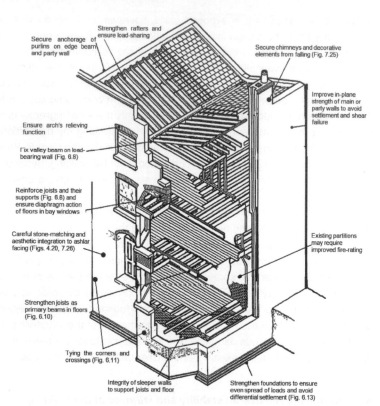

Strengthen rafters and ensure load-sharing

Secure anchorage of purlins on edge beam and party wall

Secure chimneys and decorative elements from falling (Fig. 7.25)

Improve in-plane strength of main or party walls to avoid settlement and shear failure

Ensure arch's relieving function

Fix valley beam on load-bearing wall (Fig. 6.8)

Reinforce joists and their supports (Fig. 6.8) and ensure diaphragm action of floors in bay windows

Careful stone-matching and aesthetic integration to ashlar facing (Figs. 4.20, 7.26)

Existing partitions may require improved fire-rating

Strengthen joists as primary beams in floors (Fig. 6.10)

Tying the corners and crossings (Fig. 6.11)

Integrity of sleeper walls to support joists and floor

Strengthen foundations to ensure even spread of loads and avoid differential settlement (Fig. 6.13)

5.12
Stability improvement strategies in a Georgian terraced house (after CIRIA 1994)

of Brunelleschi's dome in Santa Maria del Fiore, Florence and the restoration of its frescoes provided comfortable access for the technicians. The structure had to be carefully designed to be self-supporting to avoid shedding the loads to the church and because of its quality it was possible then to convert it into a grandstand for sporting events. An equally advanced solution was adopted more recently for another emblematic and popular building during the restoration of the dome of St Paul's Cathedral where a highly innovative rotating scaffolding was used (Gold and Knox 2005). Considering the visual effects and those of vibrations, it was eventually constructed with the capability of making eight essential movements in a very smooth manner. Unfortunately this is not the case in another famous dome, that of Hagia Sophia in Istanbul, where a scaffolding has been in place since the 1970s to allow professionals to inspect the unstable dome.

Even more delicate temporary works are required in façade retention projects. These interventions actually gut the interiors of buildings to increase commercial floor plans and they are often sanctioned by the very regulations that are expected to protect their artistic integrity by splitting it between an external and internal character (see the case of B-Listed buildings in the UK and discussion in Section 6.9).

The new structures have to be assessed for their stability during their insertion to the historic fabric and their capacity to receive the loads from the new use, in other words retention function and mechanical compatibility with the existing fabric, as well as stability and robustness. Steel systems are preferred for their prefabrication, which permits dry assembly even in tight site conditions. The result is also a structure with a more flexible location in relation to the envelope (retained façade), rapid assembly and lighter yet robust buttressing function (Bussell *et al.* 2003).

The stability of these temporary works must be guaranteed as new construction takes place, so they often need to be designed as independent from the potential bracing from the existing building. There are issues resulting from the operation (demolition, transport of materials, archaeological investigation) but also the temporary nature of the works (instability to wind loads in tall buildings where the works are located externally). The process of insertion, then, should effectively tie into the façade with a range of connections chosen for the type of load transfer or to provide temporary stability but should also offer a degree of protection for the façade and retained fabric (Bussell *et al.* 2003; Highfield 1991). Such protection should extend to weatherproofing and ensure the water-tightness of the fabric, as well as the integrity of important non-structural or decorative elements. Unnecessary movement must always be monitored during this delicate phase and also the next stages.

Following completion, there is the delicate process of transfer of loads between the new and existing fabric, and access that permits control is paramount. Eventually, once in service, the two structures must perform in harmony. The analysis of the combined structural scheme must consider the complex effect of offsets, eccentricities and out-of-plumb between the new frame and the existing walls it is built within. Anchorage and supporting strategies must be reliable for the new frame, but the entire operation can be very intrusive so the whole issue of reversibility or authenticity cannot really apply.

Stability is provided locally (bracing existing vertical openings or new ones in the floors) or globally by controlling stiffness through new lift or stair shafts or other cores, combined with the stiffness of load-bearing masonry (that is often the case) in cellular arrangements. Where strongly incompatible schemes need to be combined, their performance can be decoupled using carefully located movement joints.

A project that attempted to respect the essence at least of the existing structural scheme is the Tate Modern gallery in London, where a new steel frame substituted the original one inside the brick clad power station. The new frame 'lined' the original supports provided bracing to the original walls as it was constructed. Façade retentions are a common practice in very large historic buildings with less significant artistic or cultural character, which are also vulnerable to commercial pressures (see, for example, most mixed-development buildings along Regent Street in London).

Some extreme and pointless, from a conservation point of view, projects have promoted retention of rooms because of their 'period' features: the Marks & Spencer department store in 5 Duke Street, Dublin had listed interiors in the second and third floor that were retained by a steel frame and grid of needle beams at the underside (Bussell *et al.* 2003). An even more absurd approach, even from a design aspect, was the retention, or rather the transport of the original Georgian boardroom into the Lloyds of London iconic hi-tech building by Richard Rogers.

Stability can also be heavily compromised by fire and, similar to modern buildings, protection depends on identifying the risks and planning strategies to prevent, control and mitigate the effects of fires (Urquhart 2007; CIBSE 2010; HS 2001, 2005). Fire safety engineering includes a combination of active (fire suppression systems and fire alarms, which may be visually intrusive) and passive (fire and smoke barriers, space separation, fire prevention through minimising ignition sources, etc.) fire protection approaches. The latter also involve building design, often expressed through compartmentation: fire-resistance rated walls and floors form compartments that perform a separating function during fire exposure. Such planning must preclude integrity failure due to cracks, holes or openings that permit penetration of hot gases or even the flames. Moreover, the insulation of the non-exposed members must be prevented from reaching ignition point.

Such solutions may be less critical to masonry buildings as they often follow a cellular arrangement defined by load-bearing walls, in which case lighter partitions and especially ceilings might require improved fire-rating in order to guarantee the integrity of the scheme. Exposed frame members (steel, timber and to a lesser extent concrete) are more vulnerable, as alterations and repairs to improve their protection could affect their architectural character.

The dynamics of fire and accurate knowledge of how it generates progressive failure in a building through numerical or experimental simulation is complex. Depending on the importance or the character of the building, this analysis might be essential for the understanding of the exact exposure and a refined planning of the active and passive mitigating measures required.

Other design issues concern human behaviour during fires (mainly orientation during evacuation) and adaptation to current regulations. Introducing inflammable fabric, retardants and barriers (often in concrete) can have a crucial effect on the character of a building. Provision of sufficient escape routes can also be invasive to original space layouts and innovative strategies can be sought, as was the case with the pressurised escape routes in the Leicester School of Engineering (see discussion in Section 6.6).

A similar mixture of active and passive measures is considered in seismic protection strategies. Preventing the seismic energy from reaching the fabric (active approach) should provide dissipation of the energy transmitted by isolating the building from the ground and restoring the connection through dampers (Teflon pads or hydraulic jacks) that absorb a large part of the movement (as in the case of the Los Angeles City Hall). Alternatively, interventions can be planned not specifically to strengthen the building but primarily to increase its modal eigen-periods so that it does not coincide with the zone of major amplification of the acceleration and therefore reduce deformations (Croci 2001).

Improving the strength and stiffness of the building is the usual passive approach that is often sufficient for medium to small buildings. These initially affect individual members (walls, columns, floors, roofs) and the techniques explained in the previous section (Fig. 5.12) and the following ones (arches, foundations) are also expected to improve ductility and redistribution of loads. Improvements to the structural scheme should consider the overall stiffness of the building by carefully selecting areas that will combine to resist torsion from offset lateral loads or shed stresses away from critical or vulnerable elements. Bracing of intersecting walls is also essential and should be tied into cellular arrangements (whether load-bearing or infill in frames).

Absolute stiffness by strong binder injections to a masonry wall core or rigid connections (plates, welds) in concrete or steel frames may have adverse effects during very large induced displacements. External reinforcement with a degree of ductility is increasingly preferred, as it is expected to behave as a ductile RC member. Fibre-reinforcement (FRP) embedded in plates or fabric (Fig. 5.6) combine high strength with a minimum intrusion to the original structure.

5.2.7 Foundations

Intervening in the substructure of a historic building deals with the strengthening of weak or under-performing foundations or the extension of underground structures for new uses. Most of these operations can be quite expensive and delicate so deciding on the relevant techniques needs careful planning. Specialist designers or contractors are usually required as the situations, even after using detailed knowledge from soil investigation, may be complex and demand adaptations within limited timescales and budget.

Active approaches to strengthening existing foundations include underpinning with stronger material or bored piles. It is important that either solution covers the entire length of the footings in order to avoid settlement due to differential stiffness and response to ground conditions. Underpinning is the most straightforward solution in the conservation of a historic building and may be needed where additional loads result from the superstructure or the foundation is damaged. The sequence as described in specialist technical books (Beckmann 2004; CIRIA 1994; Alvarez and Gonzalez 1994) involves adding short reinforced pads underneath the original foundation progressively leaving undisturbed spaces in between them, which will be reinforced in a successive round. Each block is created by digging a narrow pit to reach underneath the footing and filling the alternating voids with concrete.

Another active approach when access is limited or the immediately underlying soil is not very strong is using bored piles (Fig. 5.13). The

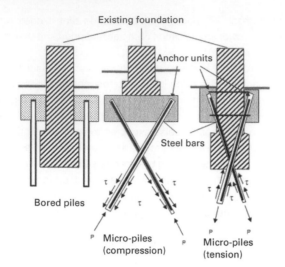

Existing foundation

Anchor units

Steel bars

Bored piles

P Micro-piles
   (compression)

P Micro-piles
   (tension)

5.13
The application of bored piles with cased steel needles and micro-piling (in tension and/or compression), fixed at a concrete pile-cap or driven through the foundation

method essentially creates a new deep support of the existing footing upon a beam fixed on two piles of 350–500 mm diameter bored into the ground on either side of the wall. Access is required on either side and the beam (concrete or encased steel) must be carefully inserted through a 'tunnel'. In this scheme, the wall 'spans' the top of the pile-caps and the loads are spread on a wider base and transferred to stronger soil strata or anchored by friction to the ground.

A solution to stitch the foundation to sound ground in confined access conditions is to insert micro-piling or mini-piling (Beckmann 2004). Steel tubes are drilled diagonally at the base of a wall (Fig. 5.13) through the footing until strong ground is reached. Once the detritus has been flushed out of the tubes using bentonite, steel reinforcement (*pali Radice*) or threaded-bars as the main load-bearing elements (like in SAS or Tubfix piles) are inserted and then concrete grout is poured in, creating a pile 170–250 mm in diameter. Micro-piles can be under compression, tension or both, according to the load they transfer through skin friction to the surrounding ground.

Applications to delicate historic structures can be seen in Ponte-vecchio in Florence and the bell tower of San Martino in Burano, Venice (Vanucchi 2007), or the stabilisation of ruins and the ground before a major intervention such as the Kolumba Museum and Neuesmuseum (Dietz and Schürmann 2006). Other advanced strategies for the control of complex settlement patterns include control piles, as in the Cathedral of Mexico City.

New foundations may also need to be provided as in the case of the settling piers supporting the crossing tower of York Minster (Feilden 2003). Reinforced concrete blocks were placed in the corners of the foundation walls, spreading the loads further on 14 m² footings under each pier to reduce soil pressures. The blocks had to be linked and further braced by stainless steel post-tensioned rods.

Passive, engineered methods address the problem directly, by trying to cancel the effect of soil failure or reduce the loads transmitted to the ground. Solutions such as retaining walls, control of the water table or techniques of soil stabilisation and remediation prevent soil failure or may even increase its strength. In the case of the Tower of Pisa, suffering from accelerated settlement in the twentieth century, the solution of 'underexcavation' at the back side created carefully located voids which closed immediately forcing the tower to tilt in the opposite direction to arrest and reduce the inclination to within safe limits (Croci 2000).

Contaminated land can cause serious problems because of highly irregular soil capacity or the presence of chemicals that will affect the durability of new concrete foundations. Solutions to improve the foundation capacity of the soil include dynamic compaction (achieving up to 0.5 m compaction of the level), vibrocompaction or vibroreplacement for mixed soils, vibrodrainage or vibro-concrete columns (600 mm diameter piles). Many of the methods have to be planned carefully in relation to the programme of the intervention.

Access and temporary stabilisation of the surrounding ground are critical. To avoid large disturbance and minimise the need for extensive temporary works (which can affect the stability of the building) a series of short pits are cut alternately and its walls secured by corrugated sheet piles that allow the space underneath the footing to be filled.

Major underground operations underneath historic buildings or stations will be discussed in detail, together with new buildings within existing ones, in Section 6.9. The sequence of these operations is critical and lead-in times have to be planned carefully. Monitoring of the structure is paramount for the control of any differential settlement due to changes to soil pressures or the water table, which can, at times, even cause uplift.

5.2.8 Arches

As was mentioned earlier, historic arches and vaults are inherently very strong structures as long as thrusts are contained at their supports either by stiffness or counter-balance. As they usually do not stand alone, apart from masonry bridges, arches are part of a scheme that sheds loads according to the line of thrust and the voussoir rib that is visible offers important reinforcement during construction and service life. The arch effect above openings is sometimes beneficial for lintels that have cracked.

These conditions are important to remember as passive strengthening approaches can resolve the direct causes of a deformation by relieving the area affected through redistribution of the loads elsewhere. Reducing the loads very rarely leads to the arch returning to its original configuration. In voussoir arches over openings, the next stage would be to dismantle and rebuild the arch. The sequence at either stage is very important, especially if the fabric above is not going to be disturbed.

The voussoirs are often finished on site to match discrepancies arising from small geometric errors, distortion from inefficient contact or

5.14
Formwork in
the repair of the
voussoirs of an arch
in Istituto Centrale di
Restauro, Rome

unpredicted movements during the application of loads. Often the blocks are laid first on the springings and crown, filling carefully the spandrels and using lime mortar to adjust the joints and align the voussoirs. The formwork and scaffolding is usually an onerous part of the process and Fig. 5.14 is an example of a kinetic flexible system that allows the location of the blocks to be adjusted and minimises disruption and costs, if used in other cases.

Some more irreversible, even aggressive solutions include the stitching of arched openings (usually in rendered and moulded masonry) with reinforcement bars and resin or hanging the opening from a beam or RC lintel. Although cost-effective in certain conditions, the rigidity they provide might be incompatible with the surrounding masonry and could alter irreparably the structural and architectural function of the arch.

Another straightforward solution is the provision of a tie at the spandrels. Although a historically familiar solution (many arches and vaults actually had bars spanning their opening as part of the original project), current technology can provide better and more compatible methods to resist the thrust. Pre-stressing is sometimes applied above the arch or very carefully along the extrados to increase the compressive stresses and eventually its strength against bending forces. Such a delicate process was applied in the main transverse arches of the rubble masonry vaults of the Masino Castle.

An interesting type of arch is what is called a basement arch, found in external staircases in the Georgian houses in Edinburgh, where access from the street is required above a basement, sometimes 2-storeys deep. This staircase is formed by two tiers of voussoir arches which, in turn, support the steps and landing, all formed from long boards or plates of sandstone. The thin arches can distort or spread if the ground on the street face of the basement is not contained. Repair is often simple as the entire structure can be dismantled and the damaged voussoirs substituted, using a plain centring, often in plywood, following the profile of the arch.

Rubble composite arches are the main load-bearing element of a bridge but their function is combined with the spandrel fill and walls. Repair and conservation, therefore, might need to address the combined effect of these elements. Some simplified theories exist to trace the thrust line (Heyman 1966), which inform other methods that aim to assess efficiently the very large number of such structures. The analysis inevitably becomes more complicated once the spandrel wall and interaction with fill are considered (see Chapter 2).

Problems with the spandrel fill often require removal of the rubble infill, waterproofing of the extrados of the barrel and relocation of new and well-graded and compacted material. Strengthening of the structure may be carried out simply on the extrados by embedding FRP bands and improving the bond of the arch ring with deep ribs often found on the spandrels and any permanent decking. Grouting, repointing or substitution of damaged voussoirs are also reversible solutions that do not alter the original scheme and flexibility of the arch rings, while sprayed concrete jacketing should be avoided. Such interventions do not affect the historic and architectural values of these bridges, which are often very important, but the need to accommodate modern traffic loads (McKibbins et al. 2006), which will undoubtedly have increased, will introduce practicalities that challenge conservation principles.

In heavily distorted arch rings, relocation of the weight in the spandrel fill may cause some re-adjustment but it has to be applied very carefully, monitoring the stability of the bridge and controlling the movement of the line of thrust at every stage. If voussoirs have to be substituted, often the new blocks will have to follow the new distorted void or be adapted on site by the masons, especially to ensure full contact and compatibility between new and old ones.

Other solutions include pre-stressing of the arch rings or even complete reconstruction, as happened with the historic bridges of Verona (Ponte di Castelvecchio, 1951, and Ponte di Pietra, 1959) after the Second World War, which were faithfully rebuilt following the same masonry forms. Further problems can occur from scour at the foundations. Where deviating footings are not provided in the original design, the pier can be secured by underpinning and piling, which are not expected to alter the character.

5.2.9 Vaults and shells

As was seen in Chapter 2, the main difference between these structures and arches lies in their two- or three-dimensional behaviour, depending on whether the thin shell or the substantial spandrels (when filled or stiffened) prevail in the load distribution. Repairs and strengthening should address the key structural behaviour accordingly.

Regarding the shell, the design constraints (finish of the visible intrados) have a bearing on all interventions at the extrados, which usually shows the rubble face of the vault. In the case of the heavily wounded vaults of St Francis in Assisi, bands of plywood reinforced with carbon fibres were applied to the intrados together with a system of dampers to control the deformation under earthquake (see Section 6.3). The techniques, overall, are not too different structurally from those applied to arches but there is stronger need to withstand higher bending and tensile forces. The use of low hydraulic lime mortar and grout was explored in the repair of the extensive cracks at the Dormitory of Kloster Maulbronn in Germany (Jagfeld 2000). Such repairs provide sufficient flexibility for the form to accommodate to the movement of the supports and lime can self-heal, cancelling the effect of small cracks. Polymeric or cementitious resins are gradually being used less as they prove to be incompatible in terms of stiffness or their durability is poor or they attract salts.

Coating the exposed extrados can reduce problems occurring from water entering through a faulty roof. Water can eventually reach the spandrel fill and cause damp, and replacing the rubble fill and then applying a protective coat can be an effective prevention (Durham Cathedral, Burgos Cathedral). In this process, it is important to allow a sufficient period of time for the area to dry, a process that can take a very long time and should be carried out under controlled conditions. Depending on the extent of the problem, an independent protective shed might have to be built around the monument, as occurred in the case of the barrel vaults of Rosslyn Chapel outside Edinburgh.

Thrust containment is often an effective passive solution and application can concentrate on the buttresses in the case of large-scale vaults. In Vitoria Cathedral, in Spain, the upper tiers of the flying buttresses were reinforced with steel bars at their extrados so that they could develop in full their arch action against the thrust of the nave vaults, eventually 'freezing' their heavy deformations.

The repair of ribs significantly affects the structural scheme of vaults in specific types, as the bond between the shell and the ribs is not always a reliable one. In some cases the rib is a substantial element that can be accessed from the extrados as happens in the English fan vaults. Reinforcement might not always make a difference or it might need to be applied

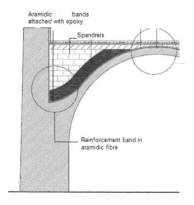

5.15
Application of aramidic reinforced bands at the extrados of a barrel vault, where they are most effective and cause less visual intrusion (courtesy Dr Alberto Viskovic)

to a deeper part of the section and be intrusive. Depending on the bond, the rib could be reconstructed.

In types like the Spätgothik ribbed vaults in Germany the ribs become a true framework where the shell is eventually an infill. In the case of St Georg in Nördlingen, Germany, whole vaults had to be lifted in the organ gallery to repair their deformation before a new organ was assembled (Barthel *et al.* 2006). The ribs were quite effective in channelling the loads of the form into a 'strutted frame' and careful analysis with FE models detected the load distribution both in the original configuration and during uplift. This enabled a controlled development of cracks in the ribs and the brickwork shells, which were repaired with lead (when small in size) or, where they appeared between the voussoirs, required these to be replaced.

Another interesting area of interventions includes rectifying past design mistakes, as occurred in Vitoria Cathedral and Beauvais Cathedral, whose collapse in 1264 signalled the end of exploration of Gothic structures into height. The reconstruction of Frauenkirche eventually faced a similar dilemma (see Section 6.4) when the engineers realised the collapse after Dresden's bombardment in 1944 was precipitated by the dome already leaning due to insufficient bracing at its base.

The strengthening of early RC shells requires a different approach, aiming to improve the bond with the reinforcement or the monolithic character of concrete, while some other durability problems can occur, such as SAR or accelerated carbonisation. Most of the techniques to repair concrete also apply here, such as cementitious or polymeric injections, shotcrete, reinforcement with metal plates or FRP bands, etc. The issue here, however, is that concrete is exposed in many of these structures and the finish forms an integral part of the character (in terms of texture, colour, etc.). Matching of the mix to be applied is paramount and the choice of the finest parts, such as cement and sand, as well as the proportion of water, controls this finish.

The examples of structures damaged in warfare are very interesting and their response is very similar to the shells in Gothic cathedrals (Gilman 1920). In Spain, the vaults at the Zarzuela Hippodrome (engineer E. Torroja, 1935) and the Fourth Water Reservoir of Oviedo (I. Sánchez del Río, 1928) were pierced by shell fire during the Civil War but their form was not affected. Zarzuela was repaired effectively after the war and later, in 2008, the long-term cracks from the bending action of the vault were carefully grouted and the pockets hosting the pre-stressing tendons were refurbished and waterproofed again.

Finally, the case of fragmentary arches and vaults found in ruins should also be noted. They are usually conserved in Roman (Villa Adriana or Villa dei Quintilli, both outside Rome) or medieval remains (Dirleton or Cadzow castles in Scotland) as evocative images of the technical achievements of their periods. Their stability is due to their over-engineered, hyperstatic design that inevitably presents sufficient redundancies, so that alternative load paths can be formed. Reintegration to the complete structural form occurs rarely, often soon after the collapse of a vault (St Francis in Assisi, Soissons Cathedral or Sant'Angelo dei Lombardi – see Giuffré (1988)). The strengthening then aims to maintain the image of the fragile equilibrium and the arch or vault is strengthened along the prevailing line of thrust. In extreme cases of ruined vaults where their structural integrity has been completely compromised, the whole vault can be reinforced externally and suspended from a light metal frame, as occurred in Cadzow Castle by Historic Scotland.

Chapter 6

Design interventions

Conservation is discussed as a design process whether it involves preservation of the fabric or an update of the building to contemporary needs or standards. The structural aspects in an intervention present equally interesting challenges that move beyond the mere application of strengthening techniques or the correct use of building technology. The main aspects have been categorised here as areas of interventions with specific structural types or technical problems that have profound effects on the structure of the historic building.

The theoretical and technical aspects covered so far usually inform a designer's choices and the following discussion and case studies guide the reader to explore the creative part of the design process and how it balances with ethical principles and technical methods.

The discussion also reflects the new priorities that structural design not just serves but actually uses to enrich architecture, such as the expression of daylight, tectonics, relationships between exterior and interior, etc. Above all, it should be remembered that all these interventions show the importance of good-quality engineering and architecture even in conservation projects, and they reflect the complexity in the role of structural engineer today.

In addition, the case studies represent the current attitude of the engineer, beyond the aim of the intervention itself and often using a scientific process in the design of structures, similar in nature to the process of architectural conservation proper. The areas chosen include new space roof enclosure (requalification of space) and protection and presentation of ruins, typical problems where a space enclosure is added and loads are directed away from the monument. Reintegration of the structural fabric or complete repristination projects are presented, where there is a complete, but sometimes controversial, respect for the original structural character. The very interesting issues of correction of past design mistakes or upgrade to current standards also offer technical challenges. Finally, the typical problems of inserting a new building with or underneath the historic fabric are discussed and criticised.

In each area, a specific structural design aspect is discussed in the main case studies. The historic context and architectural and historic character of the original building are presented, together with the technical conditions of the site. The intentions of the architect or the client are then presented and discussed alongside the available options for the structural solution.

The key aspects of the project are finally summarised and a critique is offered on the architectural and conservation effect on the historic fabric. A series of questions is given to the reader to validate the solution and assess its effect on the preservation and presentation of the original fabric. Key dimensions, conditions and loads are given in the Gazetteer. The reader is encouraged to make small physical or analytical models to test the structural strategy and commercial software (Oasys, Robot, etc.) or freeware (Easy Statics) can be used. Occasionally, structural solutions depend on environmental performance and suitable daylight studies can be performed in most CAD software or dedicated programs (Ecotect).

6.1 New space enclosure

Space roofs over open internal spaces or atria in a historic building are designed to requalify, often radically, the original design by providing more enclosed areas or rearranging access to the building. This is an area where the creative element drives the project and their strong visual character questions the limits of the intervention towards the conservation of the values of the fabric. Heavy demands or alterations to the structural character and scheme of the original fabric may result from the large thrusts that develop, while the construction process must be carefully planned.

Especially in museums, roofed courtyards offer more space for exhibition and public activities. Control of daylight and environmental conditions becomes a design consideration equally important to the efficiency of the structural form, as in space frames and gridshells designed in many historic museums (Greenwich, the Louvre, Hamburgmuseum, British Museum). Such museums are often characterised by classical proportions in the distribution of the openings, a strong horizontal linearity and stylistic unity, so a key design intent is to minimise visual intrusion and use the lightness, formal variety or patterns of space structures to architecturally enhance the presentation of the monument.

In the cases discussed here, the design of the new roof did not require restoration of the fabric as part of the design. It is, rather, an example of composition of two elements, where the original building may even compromise the success of the new roof, which has a strong tectonic character. Light becomes the new 'material' that the roof brings to the architectural character of the original fabric and the effect of its quality, together with the shadows from the structure, on the architectural character of the original must be assessed. Design in such projects often presents the ideal condition of being driven by a single engineer (or a group), which safeguards continuity of the idea and coherence in practice and communication with the rest of the design and construction team. This is often a fundamental argument for the success of the delicate operations that involve historic buildings and demand attention to detail.

6.1.1 Design process and form

The technical challenges in the design of the new structure start with the choice of the form. It should not only align with the rhythm of the underlying building but also accommodate the architecture of the existing roofs and the layout of the courtyard. The type of structure (usually space frame or a lattice) should minimise the forces on the members and ensure that minimum or no loads are applied to the existing fabric. Space frame design presents specific technical issues that are presented in these sections in more detail.

Strength in a three-dimensional structural roof usually aims to reduce bending stresses on members and is often achieved by an iterative form-finding process. This can lead to single-layer lattices (space grids) or space frames where bending is resisted by a combination of ribs, localised trusses and ties. The daylight conditions in the courtyard are often controlled by lightness of fabric, which is resolved by choice in form (single or double layer), transparency (type of glass or louvres), or density and size of members (reducing the casting of shadows).

Where space permits, a single curvature shell might be enough. In Hays Galleria in London, the space between two adjacent warehouses of 1856 was of constant span, therefore during the regeneration of this wharf Twigg Brown Architects were able to add a steel barrel vault, with hints to Victorian railway stations or the Vittorio Emanuele Gallery in Milan. This was highlighted by the use of a regularly spaced and independent frame with cellular arched ribs which, however, did not follow a contemporary tectonic simplicity but carried too strong historicist references.

In most cases, such forms have to span in two ways in order to share the loads more effectively between the members, so the primary members do not have to follow a single direction. Space frames can be strongly hierarchical structures with a dense layering of primary and secondary load-bearing elements, ties, joints, ancillary components, etc. and this becomes a primary consideration (Neptune Court, Greenwich; Richelieu Wing, the Louvre). In the case of gridshells, many of them are defined by a triangular mesh, resulting in a directionless structure. The load-bearing capacity, then, is achieved by a surface-active form, that allows mainly axial compressive forces to develop for the significant load combinations (British Museum, Hamburgmuseum).

A gridshell can provide an ideal solution for such conditions and to cover slightly offset or very irregular plans. The form-finding process potentially provides the desired lightness without excessively highlighting any primary structural members that may be required to accommodate high force variations that often result due to such irregularities. Essentially such shells need to be triangulated to give stability and the essential additional diagonals in the mesh are provided by pretensioned cables. The design process and technology are highlighted in the new roof for the L-shaped courtyard in Hamburg Museum.

Hamburgmuseum

Very lightweight single-glazed gridshell over
L-shaped courtyard

The original 1922 building is an RC frame, clad in red and black brickwork. The L-shaped courtyard was originally meant to be roofed (Fig. 6.1). The non-ornamental treatment of the internal elevations revives Hanseatic architectural styles.

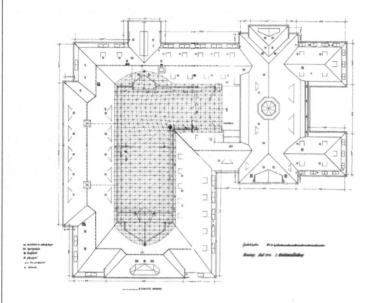

6.1
Hamburgmuseum: plan of the L-shaped courtyard (copyright: gmp von Gerkan, Marg and Partners Architects)

The project aimed to extend the exhibition space into the courtyard by somehow completing the original design and providing the environmental conditions for conservation and presentation. Its formal architecture and the rooflines required a plain and fully transparent enclosure.

This conditioned the layout of the roof and choice of materials for the structure (steel offers maximum strength-to-weight ratio) and glass (avoid heavy lamination or double glazing; low demands in thermal comfort), as well as the supports (reduce thrusts and thermal movement). The weak daylight comes in from a relatively low angle. The gable at the end of the long arm needs to be included in the form design. The construction programme should consider limited access to site and economy.

A very lightweight single-layer gridshell was chosen (Fig. 6.2), partially supported on the original RC structure, which can take moderate loads. An edge beam was provided for support, propped only at a few points on the existing RC slab (von Gerkan 1991).

6.2
Close-up of the
single-layer
gridshell over the
gables (copyright:
gmp von Gerkan,
Marg and Partners
Architects and Hans
Georg Esch)

A junction of two barrel vaults forms the shell, with shallow catenary profiles to minimise bending. A rectangular grid was chosen, that accommodates changes in curvature at the angles, clad in single glazing to minimise loads. The construction was standardised with easily available components (bars 1 m long). Stiffness was combined with lightness using pre-stressed cables (at diagonals) and spoked wheel ties at junctions.

The intervention did not pick up any stylistic or structural elements of the original in particular. It completes the act of the original project in a contemporary way. Light becomes a key material and its moderation is expected to highlight brick texture and historic stylistic values of the elevation. Space structures often suffer from a 'dated' feel but lightness here results in a gentle enclosure rather than an imposing new element.

Appraisal of the solution
Explore the lightness of the structure by using a physical or analytical model of four bays (4.8 m) to assess:

- how vertical load and thrusts change with the shape of the shell (proportions, catenary) and the type of glazing (single or double);
- the effect of ties and pre-stress on longitudinal stability and strength;
- the bracing effect of pre-stressed spoked wheels on the thrusts;
- the effect of daylight on the original architecture (orthogonal or triangular pattern, size of bars).

6.1.2 Structural layout and materiality
Due to their size, space structures may have substantial self-weight and be subject to high wind pressures, large movement, etc. Sizes and density of the members need to be controlled with respect to the architectural composition of the existing building. Moreover, the enclosing material (glass,

stretched fabric) also affects the choice of the structural system: for example, when single glazing can be used where the environmental conditions of the enclosed space are not demanding then gridshells are effective and apply very low loads.

Using primary and secondary members in space frames results in a hierarchy that might overwhelm the historic building or destabilise the composition. In Neptune Court, primary supports were formed by trusses combined with long ties, reducing the main tube section to 168 mm. Fixed joints are used to provide in-plane stiffness and create a regular rectangular grid. When used to roof a historic structure, such a layout provides directionality and elegance and careful sizing can reduce the size of secondary elements.

In space trusses, bending strength is achieved by the layering of the grid and combination of triangular and rectangular meshes, to provide in-plane stiffness. However, a single-layer grid is usually preferred as it appears less 'busy' but this requires a careful choice of the form and the glazing. Light pin joints are sufficient connections to provide stiffness through triangulation and, eventually, can be of a bespoke design which can help to accommodate the changing curvature of the planes of the triangles.

Neptune Court, National Maritime Museum, Greenwich

Minimisation of member sizes

The building was initially a naval hospital that was completed as a courtyard in 1862. Its non-unitary design has plain Georgian features and a monumental entrance at its north side, enclosing a rectangular courtyard 45 × 54 m.

The museum opened in 1937, hosting naval exhibits of various scales. The building had circulation and exhibition problems, which were resolved by utilising the courtyard. A transparent roof was chosen to provide shelter, architectural focus, uninterrupted space and exhibition conditions (ventilation and daylight).

The original fabric was assessed as strong enough to withstand new loads, so the roof would not need an independent support. However, additional exhibition space was provided by a new gallery building in the courtyard (Fig. 6.3), which would bring the shell closer to visitors at the terrace and therefore the visual impact should be minimised (Brownlie 1999). Moreover, because of the hospital's historic character, the roof should not be visible from street level. The very large spans could also cause large bending, which requires the use of either a double layer or deep members.

A cloister vault was chosen with a rather flat profile (4–5 m above the original parapets), which can exert high thrusts (Fig. 6.4). A uniform grid was chosen of 168-mm-diameter CHS tubes at 2 m centres. Strength was provided, causing minimal

6.3
Neptune Court:
plan of the roof
(Brownlie 1999)

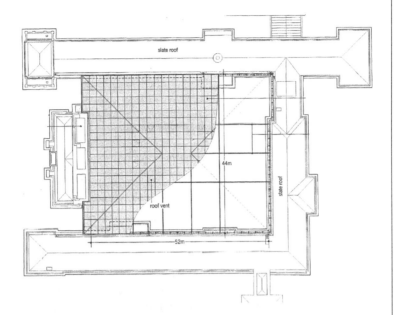

6.4
Aspect of the roof
and trussed ribs
from the terrace
(courtesy National
Maritime Museum)

visual intrusion, by 750-mm-deep trusses every second bay and ties made by pairs of 60mm Macalloy bars at 8m centres.

The entire roof is supported and braced on the original load-bearing brickwork behind the stone parapets by a concrete spreader beam, which also regularises the uneven height of the building (Fig. 6.5). The roof sits directly on vertical 139-mm-diameter CHS stilts, bolted to the spreader beam but pinned along the edge of the grid to accommodate large thermal movement.

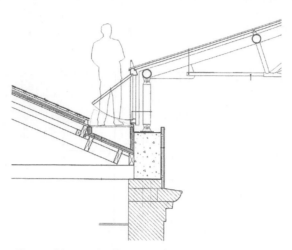

6.5
Detail of the
connection of the
roof to the aisle
of the old hospital
(Brownlie 1999)

The roof is not the lightest of solutions, but it is the addition of the gallery that compromised the impression of the original courtyard and visual integration with the historic fabric. The loads and supports on the fabric are successfully accommodated by the edge beam, carefully integrated in the roof courtyard parapets. Shadows from the roof create interesting effects on neoclassical walls and strengthen visual reference.

Appraisal of the solution

- Assess the forces on a truss in the shorter span for its own weight and wind load (0.5 kN/m²) and evaluate the effect of the depth of the truss on the size of the tubes.
- Discuss other possible solutions: pointed cloister vault, a space or flat gridshell, a series of undulated Gaussian vaults.

Ties can combine with tubes to resist bending stresses and further decrease the structural depth of the spanning members. In historic settings, however, they may detract from the openness of the space and make the roof appear industrial. The solution is often to use smaller lengths, not stretching the entire span.

Steel is the preferable material for most large-span applications because of its efficient strength-to-weight ratio. At smaller scales, however, where the materiality of the original fabric has a stronger effect on the character of the building, timber has been successfully applied, as in the new roof for the Orangery at Chiddingstone Castle that provides space for public events (Fig. 6.6). Based on experience with the Downland or Savill gridshell, Buro Happold and timber specialists Carpenter Oak & Woodland designed a domed gridshell, using soft local chestnut for the light, double grid. Laminated, toughened glazing was fixed on top of the grid using specifically designed stainless steel fixings.

6.6
The new timber
gridshell for
the Orangery at
Chiddingstone Castle
(Images supplied with
kind permission of
Buro Happold/Robert
Greshoff)

6.1.3 Supports

Due to architectural planning, spans have to be uninterrupted so the roof is supported only along the perimeter of the courtyard. This raises two main issues, the structural strategy to contain the loads (vertical and horizontal), and the structural and visual interface with the historic fabric and geometry. Because of the geometric constraints, there are not many options for the location, which requires an extra effort from the designer.

As a consequence, the strong supports are required to contain the high thrusts resulting from self-weight or imposed loads (which are proportional to the span). Thrust and weight, however, should not be shed onto the historic fabric unless sufficient strength reserves exist. Alternatively the support system should be kept as independent and reversible as possible.

Using a spreader or edge beam helps to brace the roof, as well as spreading the thrusts more evenly along the perimeter, improving also collaboration between adjacent structural members. Accommodating such a beam is a crucial design issue as it is the immediate interface between the roof and the historic building, bridging the two visually and structurally disparate systems. Treatment as a deep cornice is a historic solution that can confer an elegant barrier (see Wren's St Paul's Cathedral). A detailed structural analysis of the reactions to these thrusts could also suggest more efficient supports that could, for example avoid the corners or minimise deflections at the centre of the roof.

Movement of such spans due to heat or moisture can be quite large and needs to be accommodated to avoid inducing stresses. Joints permitting sliding or rotation must be designed in a way that can be incorporated in the edge beam and can be a technical challenge, as in the Great Court of the British Museum (Brown 2005).

Palacio de Cibeles, Madrid

Integration of support interface

The original vast General Post Office building (1918) was transformed into the City Chambers for Madrid by Arquimatica (2011). A new enclosed patio was created in the courtyard (Galería de Cristal) over a very irregular L-plan measuring 100 × 48 m (Fig. 6.7).

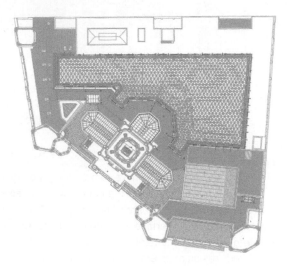

6.7
Palacio de Cibeles: Plan of the enclosed courtyard (copyright Arquimatica SPL)

The under-used courtyard was an opportunity to collect together the various civic functions planned for this emblematic building (the Mayor's Office, cultural centre, some municipal departments). The courtyard and former public areas are strongly characterised by eclectic forms, while support is provided by a robust steel frame.

The architects saw in the atrium the 'Fifth Façade' of the building and proposed a non-directional gridshell with supports only along its edges (Fig. 6.8). Its composition with the cornice at 21 m from the ground provided a continuous link with other elevations of the compound, conferring a visual unity. However, because heights of the building had been altered in its past, the level chosen left some floors above the roof.

As a result, a single-layer, double-glazed toroidal gridshell was chosen to cover spans. Its form was explored using a series of simple tests to inflate a membrane through the footprint of the edge of the courtyard (Arquimatica 2011a, 2011b). The generous budget allowed the adjustment of the shell to some very awkward corners (Fig. 6.7). Where changes in geometry were anticipated, spoked wheel ties were used to strengthen the shell and avoid increasing its thickness. The form was also dictated by the need to let daylight through the windows of the floors left outside and to avoid blockages.

6.8
View of the
new courtyard
(courtesy Palacio
de Cibeles,
Ayuntamiento de
Madrid)

The maximum possible size of glass panels set the triangular grid (Martin 2010; Porras 2011). The three-way steel lattice has a simplified support for each of the 1,920 panels along its edges and is made up of 1,034 individually fabricated 'star' nodes, each connecting six bars. All the 2,966 bars are fabricated steel tubular box beams and all steelwork was made in Spain. The glazing provided a low solar factor (16 per cent) and ventilation was controlled through some mechanical systems and natural air-flow through panels on the roof which open like vents – using a mechanism that further affected the choice of the grid bars.

The roof, weighing 500 t, is supported along a tubular edge beam which is fastened on the parapets of the roof or cornices of the building (Fig. 6.9). The 700mm tube braces the thrusts and distributes them evenly along the perimeter, shedding the loads to steel columns carefully located and concealed within the original building. A maximum of 80mm thermal movement is anticipated at the longest axis and is accommodated on sliding bearings.

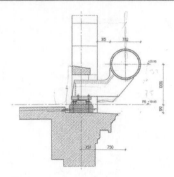

6.9
Detail of the support along the edge beam (copyright Arquimatica SPL)

The roof was assembled from prefabricated ladder beams and was erected upon precisely located props which were removed after the entire shell was in place. The whole assembly process took from December 2007 to May 2009.

The eclectic elements of the elevations were not to be interfered with as they were, visually, quite loaded, so the minimal impact of the shell was the key aspect of the project to provide a new major public roofed space. Overall, a complex geometric and mathematical model was combined with careful planning of the supports. Key to the success of the project was the central role of the architects who were able to control all specialist input, while developing their own inquiring attitude to resolving all issues that would affect the architectural expression. It is also due to a characteristic of the profession in Spain, where architects are able to certify the structural and environmental performance of their projects.

Appraisal of the solution
- Discuss limitations of alternative solutions, such as a series of arches along the short spans or a flat gridshell.
- Assess how thrusts differ when rise of the shell profile at the major spans is offset from mid-span.

6.1.4 Construction
The construction process and choice of the texture of the individual members can make a huge difference in a project, depending on the scale of the roof, its relationship to the historic fabric, daylight, etc., as will be seen in more detail in the next project.

Cour Puget, Pavilion Richelieu, The Louvre

Space frame on irregular plan
The Richelieu Pavilion (1852–57) was formed from three courts in an eclectic Empire style of architecture for the Ministry of Finance. Until 1940 it was covered by a roof on columns. The heavily ornamental exterior is contrasted with simple neoclassical lines in the courtyards.

In 1988, as part of the rearrangement of the Grand Louvre it was decided to cover the three courtyards by a transparent roof, which should provide semi-external conditions suitable for the presentation and conservation of sculptures (Fig. 6.10). Daylight conditions were fundamental for the museographic programme and the Cour Puget (middle of the three courtyards) would host eighteenth- and nineteenth-century French sculptures (Figs 6.11 and 6.12). As with many similar projects, the roof should not compromise the external view of the historic building and should apply minimum loads on the original fabric.

6.10
Plan of Cour Puget (McDowell *et al.* 1994)

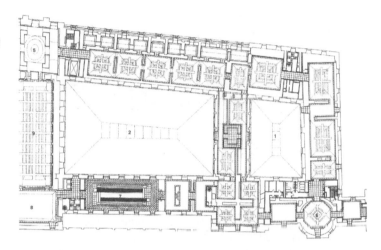

6.11
Internal view of Cour Puget

6.12
The structure of
the roof

Because of the original site conditions, the layout of the plan is trapezoid and rather asymmetric (Fig. 6.10), but elevation heights are, overall, regular. The new roof is a cloister vault designed as a space frame in a rectangular grid. The rather shallow profile was determined by the sight lines as the height of the frame structure should not be visually intrusive to visitors (McDowell *et al.* 1994).

The members had to be small in order to reduce the shadows cast in the court. This is achieved by using light transverse mullions in every second glazing bay, braced by fan-arranged ties fixed at a node clamp at mid-span, which further reduces movement and buckling length. Diagonal ties further brace every glazing bay. On top, the mullions are fixed on a spine reinforced by a light truss. The diagonal ribs were formed by tubular arches protected from wind loads by shutters. The whole roof is detached from the original parapet and is fixed on a concrete beam.

Double glazing with reflecting properties was mounted on the rectangular grid. Pre-framed sets of paralumes are fixed within the glazing, in aluminium tubes 32 mm in diameter, to provide an almost constant light throughout the courtyard.

The roof clearly has a very different structural language from the neoclassical court but it is confidently executed in a minimal steel frame that sits in front of a successful louvre system. If the eye is fixed on the ties, the solution might be compromised, but the daylight conditions treat and highlight the elegance of the elevations in a gentle and uniform manner. This, together with the right environmental conditions for the exhibits and visitors, were the other main goals of the project.

Appraisal of the solution
- Discuss limitations of alternative solutions, such as a barrel vault, or a single-layer gridshell.
- Compare with the solution used for Neptune Court.

6.1.5 Environmental performance

Environmental control and comfort also determine the choice of a structural system. Strategies for controlling natural daylight may depend on the type of glazing, density of structure, shading devices, balancing of light with the indoor areas, visitors' comfort or, in the case of the museums examined earlier, conservation of the exhibits.

The new dome at the restored Reichstag building in Berlin, for example, has a shade on the intrados of the transparent structure (Fig. 6.13). The shade rotates during the day according to the angle of the sun and is mounted on the top of the inverted cone, which structurally is connected to the steel frame of the dome. The Reichstag dome is open at the bottom at the two sheltered entrances and at the oculus at the crown, so it is fully naturally ventilated.

6.2 Protection and presentation of ruins

The conservation and presentation of archaeological ruins or heavily damaged and fragmented structures relates principally to the conditions for their in-situ exhibition, in a sort of open-air museum. In some instances it is key elements such as floor mosaics, wall paintings, important decorative

6.13
New dome, Reichstag
building, Berlin
(courtesy Deutscher
Bundestag)

or historic fragments that need to be preserved and presented and their relevant architectural context can be re-created for conservation or educational reasons. In other cases, it is the architectural space that still maintains some integrity that has to be notionally completed or to be treated as a fragment of a lost unity or container of sufficient information of the original.

Authenticity of ruined fabric as a document of architectural creation and technical ability often prevails over the architectural context and must guide the composition of a new structure, in cases where a more critical approach is taken to a conservation project. The dangers, however, of considering the philological values of the fabric as absolute can often lead to extreme or arbitrary approaches, detached from or too much attached to scientific truth. At one end of the scale, are reconstructions done with a good historicist emphasis, such as the exciting but often arbitrary restorations in Knossos, Crete by A. Evans in 1910 (Fig. 1.3). The other extreme occurs where the designer sees the need to make the archaeological evidence relevant to today as a purely creative opportunity (Sagunto Roman Theatre, by G. Grassi), diluting the fragment. Other dangers occur also where respect for the ancient document is overwhelmed by a modern architecture that purposely tries to be too close (Piazza Armerina, Italy) or too detached from the fabric (Kolumba Museum Cologne, see Section 6.8).

In technical terms, the strategies vary from the simple re-creation of spaces (usually single volumes) to large and almost neutral space enclosures that shelter a pathway that guides visitors around specific areas that are exhibited. In both approaches, the major issue is the support of the new structures and the area where the vertical and horizontal loads are discharged. The location of the supports (within or outside the site) and its interfaces becomes critical in the structural and visual interference with the historic fabric. Reintegration of the structural scheme has to consider how far to bypass the fragility of the remains, or look for a new, similar or compatible structure that is in line with the aims of the intervention and enhances the didactic potential of the ruins.

Accordingly, the materiality of the new structure has to be carefully designed, choosing whether it corresponds with the historic architecture and the appearance of the ruins, or becomes as neutral as possible. Timber, for example, is often chosen in the most successful projects due to its adaptable texture and low weight-to-strength ratio.

Interesting cases have emerged in recent years where the urban stratification is considered equally important to the historic fabric of a specific period as documentation of the interaction of society with artistic and historic values, displaying in a unique manner the cultural importance of the site (Crypta Balbi Museum in Rome, Diocletian Palace in Split). The technical aspects are challenging as layers of history and new intervention should be carefully integrated.

At the other end of this argument are cases where the ruin

is consolidated to its present precarious form either as indication of the passage of time and the effect of the elements, where this is important for the character, or even as a folly. The latter is a popular practice often associated with neo-Gothic or neo-medieval architectural trends although it existed even in the Renaissance or earlier. A recent example is White House in the Scottish Isle of Coll by William Tunnell Architecture, where the history and the character of a ruined country house were used as a catalyst for its rehabilitation, expressed by the 'freezing' of the shattered gables.

6.2.1 Re-creation of single volumes

Conceptually, when the architect attempts to re-create the volume of the original building while protecting the remaining fabric the problem is simple. Providing an impression of the original scale and architectural context is usually associated with didactic reasons and the cultural values of the ruin can be further actualised by reconnecting to modern ideas of enclosed space and urban context.

The creation of an ideal and even romanticised past for such reasons, as in the case of Knossos or Phaistos by A. Evans (Fig. 1.3), might be able to enhance the imagery of past or enigmatic civilisations but also shows the limits of such an approach. The materials and structure used there are quite aggressive and irreversible with respect to the fragile remains (concrete floors or columns, simulating the original timber structures). Wider scale reconstructions of a similar mixture of stylistic and didactic approach were attempted by Viollet-le-Duc in Carcasonne and they have damaged irreparably the fabric and cultural values.

A lighter structure is certainly preferable and should be designed in a way to apply lower loads, cause less incompatibility, be clearly distinct from the original (but not in contrast) and require a non-aggressive process of erection and dismantling. In some cases the new structure has to create an enclosed space in order to re-create the spatial values of the original and occasionally to protect delicate decorative or historic elements.

Villa Romana del Casale outside Piazza Armerina

Lightweight re-creation of original volume and geometry

A careful re-creation of original volumes along these principles was attempted in the Villa Romana del Casale outside Piazza Armerina in Sicily by Prof. Franco Minissi in 1957–58. The project was developed as one of the first examples of the critical approach to conservation (*restauro critico*) and a reaction to the irreversible and often arbitrary interventions of the recent past (Minissi 1961; Vivio 2010).

The new materials of the time, such as lightweight metal frames, large glazing or transparent polycarbonate sheets were consciously used to re-create almost the entire geometry of the Roman villa and detach as much as possible the new enclosures from the materiality of the original (Fig. 6.14). The musealisation programme was served by pathways as light decks mounted on the walls that allowed views to the exquisite mosaics of the villa (Fig. 6.15).

6.14
Lightweight metal structures in Piazza Armerina (courtesy Dr Beatrice Vivio)

6.15
Support of the metal frame on the existing walls in Piazza Armerina (Minissi 1961; Vivio 2010)

This intervention, however, has suffered from the degradation of the materials (especially the plastic roofing) and the replacement of original glazing. This degradation and alterations produced inadequate conservation conditions that often cause condensation on cold days. Moreover, the strong character of the original composition combined with these alterations disturbs the appreciation of the original forms, both because of its density and at the interfaces.

This design approach characterised all of Minissi's projects, but has also been the downfall of his work. Many of his projects have been dismantled and even here there is a proposal for a new series of enclosures maintaining, however, the current spirit, although there is strong opinion in favour of Minissi's work based on its architectural and cultural merits.

Appraisal of the solution

- If you think the current materiality is in sharp contrast to the ruined fabric, how would you choose a different material system to achieve the same spatial effect as this approach?
- Having in mind the current didactic effect, what forms would you choose to reduce confusion in the architectural re-creation?

Such re-creations always cause controversy, especially when the creative element is not of the highest quality. The visual and mechanical incompatibility of new materials has been understood by the design community and efforts are concentrated in the lightness and reversibility of the structure, as well as the interfaces. These fundamental parameters eventually condition the choice and materiality of the project, as will be seen later. The Roman Town House in Dorchester is an example of a more sympathetic approach that centred on giving the visitor the impression of the original mass and volume while viewing the floor mosaics (1996–99). The mass to stabilise the building was provided at the roof, which was carefully carried over specific bearing points (Dorset County Council 2007). The design process represents the careful steps usually taken nowadays with such delicate remains and a wide range of professionals were involved. The structure is light and robust but the effect of the roof and the tubular frame overwhelms the synthesis with the ruins. At the Roman baths in Chur, Switzerland a smaller version was explored in timber by P. Zumthor. The materiality achieves a poetic treatment and the technology successfully creates light loads over the ruins. The architecture and spatial treatment, however, overwhelms the modest ruins and makes the intervention too prominent, in a similar manner to Kolumba Museum, another project by P. Zumthor.

The problem of simply re-roofing a space is a direct result of the good conservation of the rest of the building or evidence of how the original looked not too long before the intervention. The approach of reversibility and like-for-like substitution applies here as the remaining fabric is deemed strong enough to withstand the original loads. Modern materials allow lightness to be combined with the necessary stiffness but their materiality and compatibility need to be carefully studied. Steel, as was seen earlier, does not sit in harmony with historic fabric, because of its sharpness, bland texture and large expansion rate.

Lightness was critical in the re-creation of the original vaults in St Mary's, Haddington outside Edinburgh. Vaults in Scotland are often lime-washed as they are made of rubble and new roofing was created here in the 1970s imitating the original vaults in fibreglass, this time however protected by the pitched roof above. There is a sense of technological modernity, however, which lacks the texture of plastered walls and is in an interesting contrast with the Oratorio dei 40 Martiri in Rome (del Monti 2004) where a traditional material like timber was used in the advanced technical form of glulam to create a pitched roof. The project was conceived as a means of providing a protective environment for the fragile frescos depicting scenes from the life of the Forty Martyrs. The re-creation of the original volume was thought to be the best way to present the frescos in their architectural context. Archaeological and architectural research supported the hypotheses for the height and cross-vaulted shape of the roof of the original Roman nympheum, and the visually light solution of a glulam roof (Fig. 6.16) was supported independently and combined with new brick walls, coffered in order to reduce their weight but provide the necessary stiffness.

The materiality therefore is as important as structural compatibility or durability of the enclosure. One cannot stop thinking of the powerful interpretation of the bare Roman brick wall in the Archaeological Museum of Merida by R. Moneo, which gives an impression of antiquarian authenticity. This issue is a much more subtle one in repairs or wall reintegrations. Usually materials as close to the originals as possible are recommended. Lime mortars or injections, reuse of bricks, reclaiming of timber or tiles (often of quality as high as modern ones) provide compatibility and reversibility, but should not be confused with the original fabric

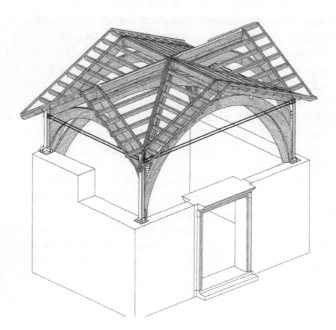

6.16
The glulam roofs at the Oratorio dei 40 Martiri in Rome (Del Monti 2004, copyright Claudia Del Monti)

or even pretend that they maintain authenticity (for example, see the case of the Odeon in Patras in Fig. 1.11).

6.2.2 Thrusts, materiality and environmental conditions in long-span shelters

A different approach to the conservation of delicate archaeological ruins is by ignoring any direct reference to the original layout and providing a shelter where the visitors are guided through specific paths. The previous technical criteria (lightness, reversibility, environmental conditions for comfort and preservation) are further affected by the size and clear span of the shelter. The supports must be located as independently as possible from the original fabric or, alternatively, the loads can be distributed onto more support points. In cases where they have to be kept away from very sensitive archaeological soil, the overall stiffness of the structure can be carefully modified or materials with high stiffness or strength-to-weight ratios, such as timber, can be chosen.

The interface with the archaeological soil conditions is equally critical in the choice of the structural type and the planning stage. The construction process should cause the minimum disturbance to the site and, therefore, the system chosen must allow for confined storage of the elements, reduced supports and assembly without heavy plant. This is often achieved by prefabrication in units or modularisation (as in the case of the membrane roof combined with a space truss in Agrippina's Villa in Pianosa – see Tensinet 1990). Adaptability to the terrain, the excavation pit or even the configuration of the ruins also requires careful consideration of the flexibility of the system and its ability to apply more uniform loads.

The planning for the environmental conditions that guarantee the preservation of the artefacts and comfort of visitors is linked to the type and extent of the envelope. Large surfaces can cause heat loss and their insulation needs to be considered within a bioclimatic strategy of the entire building. The choice between natural and artificial lighting and the need for thermal and moisture control, affects the type of translucency and type of the envelope (glazing, modern polycarbonate panels or cladding), its layout and, ultimately, the loads the primary structure has to carry. Ventilation might require major vents or even a compartmentation of the interior according to the necessary conservation conditions. Similar considerations have to be made for fire safety.

Solid shelters are often found in Roman sites where they protect and present mosaics or other pavements or evidence of the life of their inhabitants. Small-scale sites must be designed with intimacy of the fabric in mind as the scale brings the new structure too close to the remains and can cause visual conflicts if it is overwhelming. The shelters in Fregellae, Italy (2006) attempt to evoke an 'archaeological sensibility' by employing the simplicity and robustness of glulam members to create plain and elegant roofs that direct focus to the remains of the Republican city (Romagnoli

2010). Deep girders were preferred over the series of independent domuses that span about 7 m and in Domus 11, for example, the coffered pattern was arranged in a way to create an opening that simulated the form and essential daylight function of the compluvium opening of the original atrium. Primary and secondary glulam arches roofed the baths ruins, spanning 11 m and incorporating diagonal ties within the coffers.

The simplicity of the scheme is also due to the lightness of the structure which applies low thrusts and vertical loads, eventually requiring only a few, relatively light steel columns that do not block the envelope and allow more natural daylight through the open elevations. The structure is more successful in its role to promote the appreciation of the architecture of the ruins rather than its own architecture, when compared with the intervention on a similar scale by P. Zumthor in Chur. A similar neutral, almost abstract structure was built as a shelter to protect the mosaics of the Roman villa in Veranes, outside Gijón (García García 2008). The larger volume aimed to re-create the spatial qualities with regards to the original space of the main Hall, using a light metal frame, clad in earthy colours and allowing ventilation at the contact with the ruined perimetral walls.

Subtle treatment of the shelter can be achieved even at larger scales, which require different materials or exert higher loads. The protection and presentation of the mosaics and architectural space of the ruins in the Roman villa at La Olmeda, Spain (2009) was planned as an uninterrupted enclosure that permitted their appreciation in the full-scale and evocative re-creation of some original areas (Ferrari 2009). The technical demands, apart from strength, were for simplicity of the structure and control of daylight. A series of four shallow barrel vaults (span 22.5 m, length 64.5 m) formed by a steel lattice were chosen as they require few intermediate supports (Fig. 6.17). Their careful design allowed the use of thin ties and strong edge beams to contain thrusts and their weight is shed on an RC base perimeter. The lattice's fine detailing provides an elegant contemporary relevance to the site (Fig. 6.18), while the location of the supports in only a few columns minimised their effect on the envelope, which was then clad in translucent polycarbonate panels that controlled the natural light.

It is very interesting to see how technology and architectural sensitivity to the values of the archaeological past have evolved when compared with the earlier intervention at the Roman Palace in Fishbourne, England (1968). A very similar programme was created but the lack of experience with high-quality lattices at the time led to a folded plate roof. The resulting dense arrangement of the supports and confidence only in small glazed panels created a visually obstructive envelope which, despite the good intentions, might not create the best environmental conditions for the conservation or appreciation of the mosaics.

At the other end of the scale, depending on the nature of the remains or their fragility, natural daylight might be completely blocked, creating full museum conditions (as in the Domus in Ortaglia). Here the structural system, or even the proportions, play a much less important role as it is the

6.17
Interior of the shelter in the Roman villa of La Olmeda (copyright Villa Romana de la Olmeda)

6.18
Detail of the vertical enclosure of the shelter (copyright Paredes Pedrosa Arquitectos)

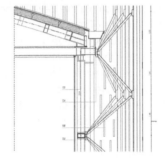

artificial light and treatment of the cladding and surfaces that support the presentation of the ruins as artefacts.

In some cases, the design of such enclosures should also provide the possibility for expansion, when the site is further uncovered. In many cases, however, a protective enclosure is required that is temporary and flexible. Depending on the conservation requirements and the environmental conditions, the structure might have to be translucent and light, as in the case of the modular membrane canopies that protect Agrippina's Villa in Pianosa or the quite complete remains of the important Temple of Apollo Epikourios in Greece (Fig. 6.19). Often the problems in such temporary shelters are their durability (30 years) and the location of supports (to withstand the high anchorage forces), as will be seen in the next section.

6.2.3 Direct interaction with the fabric: location of supports and additions

The transformation of historic buildings by adding new fabric has affected many large-scale buildings (such as palaces and castles) or urban sites

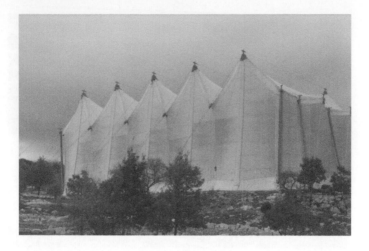

6.19
The membrane
shelter over the
temple of Apollo
Epikourios in Figalia,
Greece

throughout history. The most evocative examples are usually medieval castles or Roman structures, for example the development of Diocletian's Palace in Split into the urban nucleus of the city.

When dealing with direct design, it is usually the location of the
supports and the mechanical interaction with the original fabric that are the
technical concerns in the project. Like-for-like substitution is a good practice
that is successful only when there is detailed knowledge of the past performance of the building. Such interventions are the main priority when the
building needs to be completed in order to restore its structural integrity (as
will be seen in the next section), but often they are instigated by the need
to add new spaces or provide an integrative and permanent shelter for the
ruined fabric. If a composition of the existing architectural values with the
new, current ones is the best approach, interaction with the original scheme
of the fabric, however, might be unfeasible due to uncertainties about its
safety.

Concerning the location of the supports, the perimeter of a site is
a safe option but might not be enough to guarantee the stability of the structure. When further, intermediate supports are required, their impact on sensitive archaeological soil needs to be considered, first in terms of their effect
on the comprehension and character of the site, and then with regard to the
potential irreversible damage to the remains in the short or long term.

A new shelter was designed in Akrotiri in Santorini, Greece (Fintikakis 2004), to protect the fragile remains of the Bronze Age city buried by
volcanic eruption around 1613 BC. The necessary environmental conditions
were provided by 14 barrel vaults formed by space trusses in a north-light
layout. The system was chosen for its lightness and modularity that enabled quick assembly in the remote island conditions, and was supported on
tubular columns in random locations as the layout of the city, the ruined
houses and the visitors paths allowed. The roof was thermally insulated
by 15 cm of earth, but it is likely that irregularities in the depth of the soil,

which added weight, combined with the complexity of the supports and environmental conditions, caused the fatal collapse of part of the shelter in 2005. Following repairs and improvements in the presentation, the shelter is expected to reopen in 2012.

A similar but more open shelter was designed in the form of membrane roofs to protect the Hagar Qim and Mnajdra megalithic temples in Malta (Fig. 6.20). The form was defined by the need to restore the astronomical alignment of the temples, and the light trussed arch supporting a membrane roof was considered the best solution to minimise structural impact, ensure reversibility and provide the necessary protection conditions. Biaxial cablenets combined with the membrane managed to stabilise the arches without additional and visibly intrusive cables (formTL 2011).

Location of supports can become an architectural issue in superstructures built above ruins as a means of integration at urban level or musealisation of the remains. This was the case of the Diocesan Museum of Cologne in the remains of the St Kolumba church, where a project (1999–2005) planned to display the excavated ruins and allow understanding of stratification of the site and Gothic church layout, while preserving a chapel built on the site in the 1950s. There was, however, a major addition built above the ruins to provide gallery accommodation for the extensive collections. The building added overwhelms the original structure and aims to balance the character of the site with the daylight and environmental conditions necessary to conserve the ruins, while the interface with the few standing Gothic walls was treated with the materiality of the new envelope.

RC columns were carefully located on pads and aligned with some of the original piers, as a reference (Fig. 6.21). Manipulation of the stiffness of some perimeter walls enabled transfer of loads away from critical areas. Whether other, subtler structural strategies would have been

6.20
A membrane open shelter protecting the Hagar Qim and Mnajdra megalithic temples in Malta (copyright form-TL)

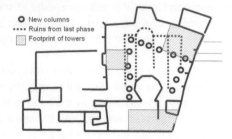

6.21
Plan of the Kolumba
Museum showing
the relation with the
Gothic remains (after
Davey 2007)

possible depended really on the intentions of the designer. Externally, the intervention appears as a rather new build, with minimal links to the history of the site, as the perforated band at the interface with the ruins shows no connection (Fig. 6.22). Internally, viewing of the ruins is successful but they are clearly treated as a museum piece detached from their space function.

This project has shown that adding an entire new building rather than a shelter over an archaeological site will always be an approach that aims to leave a strong mark for the designer's benefit. The same criticism has been levelled at other types of permanent structures designed to protect and exhibit entire buildings, ships or large-scale artefacts. It is inevitable that once these objects are treated as items disassociated from their original environment, the new architectural context is essentially arbitrary. If the designers do not have the sensitivity, knowledge or awareness of the richness of the site and its cultural context, then the solutions often give protagonism to their own design (as in the case of Ara Pacis in Rome) or force

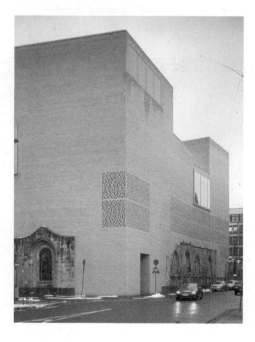

6.22
The relationship of
the new structure to
the remains of the
Gothic church (M.
Glendinning)

(almost abuse) technology to achieve a specific effect (e.g. the proposal to integrate the delicate Cutty Sark clipper into a steel and glass cradle, allowing it to be viewed as main 'exhibit' as part of a restaurant in Greenwich, London). Such solutions are in sharp contrast with other more respectful approaches, as the in-situ conservation of the remains of the platform of the Jupiter Capitoline temple within the Capitoline Museums in Rome.

Taking this approach even further, occasionally, entire buildings are created to almost engulf the delicate monument, as occurred in the Arab Baths of Baza in Granada (Ibañez 2010), the Ara Pacis museum in Rome by R. Meier or other structures with religious importance, such as the new church surrounding the hut of St Herman of Alaska on Kodiak Island.

The areas examined so far have dealt with new protective structures external to the archaeological soil with various degrees of interaction, scale or intention. However, the historical values associated with these sites appear also in layers above ground, and sciences such as medieval archaeology consider the stratification of entire buildings to be vital for the understanding of the evolution of sites, especially in urban contexts. There are technical problems of how such distinct fabric interacts or how to expose and present layers of history without disturbing the stability of the building. This also requires a location of the protective and exhibiting structure within the fabric: parts of earlier phases might need to be reconstructed or ramps might have to be added to allow access (Fori Imperiali, Rome in Fig. 6.23).

6.23
The enclosure and ramp of the 'Mercati Traianei' in Rome

Crypta Balbi Museum, Rome

Interaction between layers of history and structure

The complex history of the site represents the development of the city of Rome. At the site of the Roman Theatre of Balbo, houses, a monastery and churches developed during the degradation of the ancient city in medieval times (AAVV 2000; Lorenzetti 2000).

As part of the Archaeological Museums of Rome network, the project (2000) wanted to highlight the stratification, treating the building as the main exhibit. Located in an area of continuous urban exploration, the Guidi House was chosen and in the future it will be linked to the rest of the site. An approach of urban conservation was sought and modern materials (steel, RC) and forms were used in a dialogue with the historic architectural qualities, creating new compositions (Manacorda 2004). The strength of steel enabled pure lines and minimal insertions that did not disturb the appraisal of the relation between the Roman and later fabric.

This vocabulary permitted the successful integration of the visitors' pathway in a reversible and distinguishable manner through existing openings (Fig. 6.24). In addition, original elements found in-situ, such as fragments of the porticos, were reconstructed in their location (Fig. 6.25).

Appraisal of the solution

- Timber was shown to be a visually and mechanically compatible material in other interventions but what would have been the effect of its use in this case?
- Areas of the building had to be emptied to accommodate in-situ remains. How would you balance the damaged integrity of the medieval buildings, in visual and structural terms, with a reconstruction of a fragmented Roman form?

6.24
Crypta Balbi: access and dismountable ramps carefully located among the ruins (with kind permission of the Ministero dei Beni e Attivitá Culturali – Soprintendenza Speciale per i Beni Archeologici di Roma)

6.25
Crypta Balbi:
recomposition of
original elements
through the
stratification of the site
(with kind permission
of the Ministero dei
Beni e Attivitá Culturali
– Soprintendenza
Speciale per i Beni
Archeologici di Roma)

6.3 Reintegration of the structural fabric

Interventions in this field are the consequence of damage to the fabric, either intentional or due to inherent reasons, which fragments the architectural unity and structural integrity of a building. The problem is closer to pure conservation principles, especially those relating to artistic values where the fabric is the essential vehicle of the creative expression.

It is very important to consider that the fragments used to be part of a whole, and the decision is often whether to reconstitute them as such and restore a unity, or regard them as entities that condense or evoke the values of the original and which, therefore, can be treated as self-contained. Such an approach, with a romantic touch, has inspired a cult of ruins (Riegl 1903) and even lies behind the often chimerical quest for material authenticity in conservation (SPAB Manifesto).

The technical considerations are, therefore, the understanding of whether the fragmented fabric can be recomposed to its original unity and how lacunae, the missing parts, can be treated. As will be seen in the case studies, this could affect the envelope (if material has to be replaced) as well as the structure of a building (how reliable is the original scheme). The opposite problem is the case of architecture that incorporates spolia from previous periods (such as the late Byzantine church of Ag. Eleutherios in Athens): even when their aesthetic values are composed in a balanced manner, technically they are very irregular building blocks whose materials may function incompatibly because of their different properties and unsuitable location (a cornice functioning as a window jamb, for example).

One of the most usual problems is the substitution of fabric either for aesthetic and durability reasons, or for restoring a structural stability to the building. This could be partial, on specific areas chosen according to the problem or didactic programme of the conservation project, but often addresses entire parts of the building that went missing due to calamities not long before the intervention.

6.3.1 Appearance and durability

This is often considered as the most straightforward application of conservation technology. Damaged material can be directly substituted by a new one and as long it comes from the same quarry or is produced by the same techniques, authenticity is served.

Confusion commonly results when such simple repairs are identified with conservation or even restoration, pretending that historic or artistic values are about the fabric alone. Unfortunately this conception is shared among engineering professionals as well, who would simplify structural problems by merely substituting original members for their function, or even worse, for their appearance, for example when recommending prefabricated stone arches to bypass the expensive formwork and carving stone voussoirs that a proper rebuilding requires.

Most importantly, material should be chosen according to physical and mechanical purposes rather than authenticity or similarity to the original. This is very often the case for facing stone, where durability is much more important than short-term matching in colour and texture. Compatibility with the adjacent intact blocks in terms of its response to thermal movement, freeze/thaw cycles, environmental pollution, discoloration, etc. must be carefully specified and ideally tested in experienced laboratories (such as the British Geological Survey).

Conservation of neoclassical stonework in Edinburgh

Ashlar stonework characterises most of the New Town in Edinburgh, built in various stages after 1780s in an attempt to modernise urban life in the city and bring it into line with the institutional British architecture of the time.

Stone is used in carefully dressed blocks at the elevation of buildings, for structural functions (such as columns or mullions in bay windows), but also in ornamental elements such as cornices. The aesthetic purity and geometric elegance of the city is compromised by natural or atmospheric degradation of the stone as well as deterioration of the pins that hold blocks in their place. There have been serious accidents recently that forced the City Council to obtain special statutory powers to oblige building owners to look after their property and perform repairs promptly.

Beyond the enforcement, it is essential in such conditions to actively inform the process, and the best manner is to collect and disseminate good practice among architects and contractors, while at the same time nurturing the industry. A major problem is the lack of specialist stonemasons, therefore education becomes paramount. A method in this direction is the creation of training models, such as those prepared at the University of Edinburgh (Fig. 6.26).

6.26
Model of a typical Georgian chimney in Edinburgh (fabricated by Adam Neep, Emma Garland, Lee Kynoch; University of Edinburgh – photo by Rachel Travers)

A further typical problem is the contrast between new and original stone. Stone ages relatively slowly and the different tones really take years to match. The effect on the elevations is very unsatisfactory as it is patchy (Fig. 6.27). Technical solutions that include artificial ageing are expensive and matching the degraded tone may cause problems in the long term.

6.27
Effect of stone conservation in a tenement in Montgomery Street, Edinburgh

Appraisal of the solution
How would you use models to assess:

- effect of gravity;
- long setting times of limes;
- bottom-up process;
- bond with rubble back;
- durability;
- ageing of materials?

Interesting cases in substituting original stones often result from war destruction or cancelling the memory of uncomfortable periods. *Damnatio memoriae* is practised nowadays in a more subtle manner as was the case of Nazi buildings in Berlin, where the stone cladding that displayed hated insignia was flipped over and made to disappear. This is clearly an example where authenticity is not desired. Similarly, the almost intact fabric of the Congress Hall in the Nazi Party Rally Grounds in Nuremberg (Zeppelinfeld) was architecturally pierced by a new deconstructivist structure when it was transformed into a Nazi Documentation Centre (Doku-Zentrum) by Günther Domenig in 1998 (see images at www.kubiss.de/kulturreferat/ reichsparteitagsgelaende/englisch/dokuzentrum.htm).

Correct differentiation in stone, therefore, needs careful planning and a direct solution is often to reuse similar material from the site (stone, ancient bricks). When it is not the subject of a delicate recomposition of original fabric (as in the Akropolis monuments) material that has not been stored away from moisture, vegetation, direct contact with materials attracting salts (such as concrete), etc. should not be selected until it is treated. In this sense it might be wiser and more truthful to use clearly distinct materials such as modern brick (as in the case of the Baths of Sikyon, transformed to a museum in 1939 or the Roman Odeon of Patras – Fig. 1.11) which avoids such problems and allows their durability and strength to be more easily controlled.

Some materials, such as concrete or resins, were also popular for such repairs until the 1980s. For reasons of incompatibility and adverse long-term behaviour and degradation that were explained in previous chapters they should be avoided. It has to be noted that the industry has advanced significantly to the degree that concrete is often named as 'artificial stone' because it is now possible to produce it with fine textures and colours. Concrete is considered as a designer's material and the achievement of these properties needs knowledge and very careful planning of the process, which often requires direct involvement of the architect right up to the production stage (Bennett 2001). Proper artificial stone products, on the other hand, often come ready in blocks and specifying them, once again, needs a careful study of their long-term performance and compatibility with the adjacent stones.

Façades and envelopes in historic modernist architecture go beyond the mere specification of correct materials as they are systems that depend on the availability of components from the industry and relate heavily with secondary structure in the building. They will be studied in Section 6.7.

6.3.2 Structural and aesthetic reintegration

As in the previous case, the problem moves beyond structural strengthening and repair, to address recomposing the original material or using traditional techniques in order to restore a degree of structural integrity for stability and educational reasons.

The approach relates heavily to the nature of the fabric. Theoretically the case is easier for dry construction (as always), where the material is not permanently bonded with mortar, as such structures can be re-assembled. The authenticity is difficult to ascertain as the fabric could have been disturbed without leaving much sign and the lack of organic material in the bond can make dating complicated.

The case of stabilising and restoring prehistoric drystone structures such as the Iron Age brochs in Scotland (roughly 250 BC to AD 250) poses problems from the fact that their architecture is 'anonymous', i.e. it is difficult to inscribe in a historic/evolutionary process. Progress and good practice are complicated to establish (Fig. 6.28), especially where the

6.28
The irregular gneiss blocks of Clachtoll broch, Sutherland

material culture and findings are not very rich. Occasionally, patterns in the assembly of the stonework can be identified and the construction process can be understood, especially at the delicate edge of a ruin from where the degradation front can progress. Restoring the integrity depends mainly on archaeological documentation that determines loose material in need of fastening, which is what often drives such projects. The stability of the original broch depended also on the continuity of the ring, so measures to complete the stone envelope can safeguard stability using distinct material between old and new. The field with regard to brochs has been characterised by simple replacement of stones or filling of voids, but uniform treatment of the voids with different or dressed stone should be followed as part of an architectural project, aiming to create a modern culture for brochs and to structurally brace the ruins, improving their integrity.

Acropolis monuments, Athens

Recomposition of structural elements and repair of past damage

The high quality of the monuments witnesses the advancements of their era and the current works are setting the mark in conservation of this type. The structures (essentially post-and-beam with columns built of drums and the entablatures assembled with iron pins) have behaved remarkably during their lifetime as damage has been caused mainly by human action such as the Venetian bombardment (1640) and the ill-conceived works of Balanos (1896–1902, 1923–33).

The current programme is guided by anastylosis (relocation of original material) in the spirit of the Venice Charter (reversibility, respect for authenticity). The intervention is restricted to the minimum necessary to respect and preserve the classical structural scheme and function of the members. Stability is restored only on those members or areas where it is compromised and where their legibility as part of a missing whole will be improved.

Interventions range from integrating small fragments to completing columns and beams (Fig. 6.29) or reassembling entire buildings, such as the Athena Nike temple. In this process, the damaged areas are dismantled and fillings or replacements are made after careful geometric study (Fig. 6.30; see also Karanassos (2006) and Tanoulas (2006)). Pentelic marble is used for its compatibility and, once the complex interfaces are assessed, the blocks are joined with white cement and high strength titanium clamps or dowels that minimise the intrusion of holes.

Overall stability is simple and efficient, with inherent high redundancies, but full control of a structure made of discrete jointed stone blocks subject to seismic action is impossible. The eigenmodes cause complex concentration of stresses (Ulm and Piau 1993), but the repair to the structure does not aim to alter the original scheme and uses existing joints as much as possible. In a limit state, failure will occur on the connecting members, not the fabric.

6.29
The new Ionian capital at the Propylaia of the Athenian Acropolis (courtesy ΤΑΠΑ –Hellenic Archaeological Receipts Fund)

6.30
Upper contact surface of a Propylaia column (Karanassos 2006, courtesy YSMA and Dr Konstantinos Karanassos)

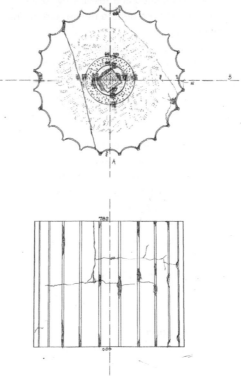

> **Appraisal of the solution**
> The effect depends on craftsmanship and careful study of each fragment, a long process with unexpected changes.
>
> - How would you summarise the main issues when introducing new marble on highly eroded interfaces?
> - What range of professionals do you think are involved in the recomposition of a column?

The same philosophy of anastylosis can apply to other types of structural elements, such as arches found in windows or vaults. The approach entails the replacement of their moulded voussoirs, which requires a delicate operation to keep the structure being supported in place. The logistics might make it more convenient to replace the entire blocks rather than the damaged parts (see, for example, the windows of the Great Hall in Stirling Castle). Arches are often very incisive elements in the character of a building. The case of the 'Gothicisation' of the vault ribs in Santa Maria Sopra Minerva, Rome by Girolamo Bianchedi (1848–55) by carving them into pointed arches is a testimony to their visual and structural importance, in line with the debate of the role of ribs in Gothic architecture around the period.

At the other end of the scale stand almost faithful re-creations of the ancient volumes with new materials, an approach that falls under the heading of repristinations. The restoration of the Arch of Titus in Rome by G. Valadier is a key early example of conservation and the associated approach is discussed in the next section.

The nature of the fabric makes a substantial difference to the extent of substitution and the distinguishability of materials. Bonded structures, such as masonry, result from a wet construction process and it is very complex to make any repair or addition reversible. Essentially it is a case for reconstruction with compatible materials as often happens in ruins, for example Villa Quintilli and Villa Adriana around Rome, or the Roman Odeon in Patras (Mallouchou-Tufano 1998). A less intrusive approach to full substitution is to address binder loss, especially when the rubble core has disintegrated. The aim to restore structural integrity can be served in a fully reversible manner by injections of superfluid lime mortars, which is in keeping with the nature of the system.

6.3.3 Structural repristination

In the case of entire structural elements, repristination of their original function might be possible without improvements. There is a subtle difference between the case of repristination of entire buildings due to the scale and approach to original structure (Frauenkirche and similar in 6.4) and the correction of past design mistakes (Fallingwater in 6.5).

Very often such interventions affect monuments of national importance that have been recently damaged. The case of Stari Most bridge, in Mostar became one of the symbols of the Civil War in the former Yugoslavia: after its destruction in 1993, a full replica was built in 2004 in the same stonework construction, reusing as much original material as possible.

St Francis Basilica in Assisi

New technologies in the reconstruction of the vaults

The Basilica combines the tomb of the saint (in the Lower Basilica) with the unique frescoes by Giotto (1296–1304) in the Upper Basilica, which is modelled as a single nave with circular ribs and pointed webs, restrained by buttresses (1239–53).

The earthquake in 1997 had a 0.18 g peak ground acceleration that was amplified 3 to 8 times in the vaults because of the narrow and tall nave. The tragic losses (four victims and parts of the frescoes) prompted the reconstruction of the damaged vaults and reinforcement of the remaining ones. The intervention was designed with respect for the original structural scheme, which was mildly updated.

Flexible yet strong ribs were added at strategic locations on the extrados of the webs (transverse arches, diagonal). Using strips of mahogany plywood as a stiff nucleus (Figs 6.31 and 6.32), compressive strength was achieved by adding glass fibres at the extrados while tension was resisted by aramidic fibres at the intrados (Croci 1998, 2000).

The vaults in the transept and the façade collapsed due to high shear stresses caused by stiff diaphragm constraints, accentuated by local effects of the spandrel fill,

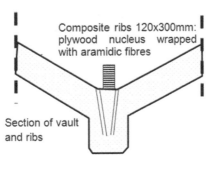

Composite ribs 120x300mm: plywood nucleus wrapped with aramidic fibres

Section of vault and ribs

6.31
Detail of reinforcement above the existing diagonal ribs of St Francis in Assisi (after Croci 1998)

6.32
The dampers at the extrados of the strengthened vaults in St Francis (courtesy Dr Alberto Viskovic and Studio Croci)

consisting of building debris. The reconstruction reused structurally only bricks at the ribs and the centring required careful geometric study of the curvature and a delicate process of assembly and removal.

Appraisal of the solution
* Do you see similarities with RC structures and how do you think the reinforcement works without altering the structural scheme?
* What issues do you see in the removal of the centring if it has to sit on capitals at the springings?

This approach can also be applied to entire structures. In principle, stonework in Renaissance and Mannerism styles was viewed ideally as a sculptural exercise, similar to the classical Greek techniques. A structure could be designed in blocks on board and carved by stone masons and theoretically it could be assembled and dismantled with the same ease. The stereotomy of the stonework in the undulated façade of San Carlo alle Quattro Fontane allowed it to be completely taken apart and replaced during conservation (Ceradini 1993). In fact, the careful study of the façade demonstrated the high quality of its execution and its original 'regola dell'arte', so the intervention actually involved doing only minor maintenance to the structure (the travertine was meticulously cleaned).

Such interventions are possible due to the nature of the structure. During the removal, however, blocks may fail due to removal of the confining forces that kept fragmented stones together, or simple damage. The repair of stone carefully carved for a specific place can be as complex as the completions taking place in the Acropolis and the debate on authenticity should focus on the originality of the proportions and the structural scheme.

6.3.4 Dismantle and reconstruct

The extreme case of monuments being dismantled to be carried away from their original location appears to have technically the same problems and is not recommended. It often happens when monuments are threatened, such as lighthouses at the edges of eroding cliffs in England (Belle Tout), the monuments of Abu Simbel or where other important buildings affect the site, such as Gaudi's Escuelas that are in the way of the expanding Sagrada Familia in Barcelona.

Ancient monuments are also sometimes recomposed in an effort to conserve and present them in museum conditions, for example the delicate Etruscan tombs in Tarquinia, the Pergamon Altar in Berlin or the Tholos in the Epidaurus museum in 1907 (Mallouchou-Tufano 1998). There are clearly educational reasons, although the legitimacy of stripping archaeological objects from their site and landscape or even moving them to another country is debatable.

Moves are usually carried out to safeguard monuments from the destruction of their site. In archaeological sites it is part of a process of unearthing earlier historical phases (e.g. the reconstruction of the Horti Farnesiani Gate after its removal due to the major excavations in Rome's Forum in 1883). However, the operation is often so aggressive that it allows contractors and clients to take shortcuts and bypass the problems with keeping the original structural scheme. The scale and logistics of such projects are not helpful either.

An urban example is the Trinity College Kirk (main phase dating from 1460) which was moved from the foot of Calton Hill in Edinburgh in 1848 during the construction of Waverly Station (Fig. 6.33). In both cases some arbitrary choices were permitted during the reassembly. Some pieces were not integrated in the Horti Farnesiani Gate. The blocks in Trinity were all numbered but were left to the elements for too long, so when a reduced version of only a transept and the choir were finally built off the Royal Mile many stones were unidentifiable and are clearly misplaced.

In general, such operations can often be so aggressive that they allow contractors and clients to take shortcuts and bypass the problems associated with keeping the original structural scheme, while in other cases 'improvements' or new structures are inevitable. The scale and logistics of such projects are not helpful either. There is a full range of attitudes that challenge or even ridicule authenticity, from the move of Palazzetto Venezia in Rome to make space for the Vittoriano (1911) to the sale and transfer of Spanish cloisters to the US (Ancient Spanish Monastery in Florida – see next section).

Finally, reference can be made to the value or even poetics of copies and casts of prominent large-scale architectural elements. Full-scale replicas usually represent only the form of a specific phase of the original, using contemporary structural technology to ensure stability (such as the proposed reconstruction of Euston Arch in London or the reconstructed box-frame medieval house as the Lace Centre outside Nottingham Castle). Immediate reconstructions also demonstrate this split between the authenticity of form and the use of current technology, as summarised by the '*dov'era com'era*' (where it was, as it was) dictum that guided the reconstruction of St Mark's Bell Tower in Venice.

Copies can also have a highly educational role in allowing a hands-on study of a historic technique. All Architecture and even Fine Art schools used to have an extensive collection of casts used in their teaching programmes. Most of them were abandoned with the advent of Modernism and one of the few remaining is being restored at the Edinburgh College of Art (which includes even direct casts from the Elgin Marbles). The success and importance of the Gothic portals created by the Centre des Monuments Historiques in Palais Chaillot in Paris are a witness to the scope of such projects. The collection of full-scale models (including, even, a representative apartment from Le Corbusier's Unité d'Habitation) at their Cité d'Architecture allows direct appreciation of the ability of construction technology to create some crucial architectural forms.

6.4 Repristination in restoration

The issue is usually the re-creation of a historic form, often recently deteriorated or destroyed, to various extents and for a variety of reasons, such as national identity or recovery of a damaged architectural unity. The difference from reintegration here is the extent of the intervention. Usually they range from reproduction of form in its entirety, to re-creation of its essence or fragments being reintegrated in the original form.

There is a difference here from cases such as St Francis in Assisi (Section 6.3) mainly because in that case there was a structural necessity to reconstruct the vaults for the stability of the monument, together with the need to heal a recent painful event. On the other hand, there is often an interesting dimension in repristinations where faults in the original design have been detected, partially leading to the failure. The critical approach sometimes results in corrections to the original scheme (as in Frauenkirche) and this aspect will be further explored in Section 6.5 which deals with design errors.

6.4.1 Faithful reproduction of a ruin

Typically, full reconstructions occur soon after a traumatic event. St Demetrios, in Salonica was immediately rebuilt after the Great Fire of 1917 also serving as a cultural focus for the reconstruction of a major city that had just become part once again of the Greek dominion. San Paolo fuori le Mura in Rome was reconstructed after a fire in 1823 explicitly avoiding any changes to the original form (Leo XII ordered '*Niuna innovazione dovrà dunque introdursi nelle forme e proporzioni architettoniche*' (Therefore, no new elements should be introduced into the architectural forms or their proportions)) highlighting a critical and conservative turning point to the architectural treatment of ancient monuments (1825–54). Many emblematic churches had to be rebuilt in Russia after the fall of Communism in 1989. All these cases maintained some fragments of the historic fabric but contemporary materials and systems were used throughout.

This was not the case in other major reconstructions, such as the bridges of Florence (Santa Trinita) and Verona (Ponte di Castelvecchio, Ponte Pietra) following destruction during the Second World War in 1944. Most of the original material that was not damaged was reused and largely the same construction processes were followed (Cecchini 2004).

Maintenance and modernisation of a structure is a continuous process that has occurred historically. Important phases in the lives of monuments have been triggered by the need to substitute an old-fashioned, failing or under-performing structure and the approach to the pre-existing forms is also evidence of the technical and integrative abilities of the times.

For example, the ground-breaking vaults of the choir in Durham Cathedral (after 1093) were replaced after 1235 and the original innovative ribs were updated to the style and techniques of the period. The roof of the façade of Palazzo Carignano in Turin was renovated by A. Bruno in 1993 using glulam beams to combine the plasticity of the Baroque forms with efficiency for the utilisation of the space (see Section 6.4.3).

The movement of monuments is often a controversial issue, as was seen in the previous section. The temple of Abu Simbel had to be relocated during the construction of the Aswan Dam in Egypt (1968), but the Axum Column in Rome was truly a trophy from the colonial times of the Italians in Ethiopia and was eventually repatriated. 'Improvements' or different structural systems are inevitable, sometimes as a result of the movement process. Detachment from the construction process or efficiencies often cause further shortcuts in the reassembly, as was seen in the problems with Trinity Aisle in Edinburgh. The Ancient Spanish Monastery (the 1140 St Bernard de Clairvaux in Sacramenia, Segovia) bought and transferred from Spain to Miami Beach in 1925 by W. R. Hearst is another example. The building was not re-assembled for 26 years and by then the original numbering had been lost, leading to an almost arbitrary reconstruction. Interestingly enough, the building is used now as a church.

Fakes and reproductions are also produced, for educational, recreational or sometimes nationalist reasons. Examples include traditional houses in expos (Pueblos Españoles in Barcelona and Gijon, Spain; Borgo Medievale, Turin; the Disney EPCOT in Florida, etc.). These are cases where the technical issues and possibilities discussed here are irrelevant; some relation between fabric and its construction process is showcased but purely on formalist terms and, eventually, is pointless as the use of current technology permits liberties and technical shortcuts to be taken.

Many of the cases in this category are reconstructions following recent traumatic events such as warfare or a natural disaster. This *dov'era, com'era* attitude is dictated by national or cultural attitudes and can be viewed according to various degrees of critical approach to past design.

Many cathedrals in the north of France were immediately rebuilt after they were deliberately targeted in the First World War in 1914 to suppress the morale of the defenders. Most of the cathedrals had their timber roofs burnt down and their vaults collapsed or pierced. Their reconstruction was immediate: Rheims (Henri Deneux, 1919–38), Soissons (Émile Brunet, 1928–37), Noyon, Senlis (Émile Brunet).

Post-war austerity and a lack of stonemasons trained in conservation and historic techniques delayed the projects, while there was not enough good quality timber and stone available. Apart from the faithful reconstruction of the stone vaults, where as much original fabric as possible was reintegrated, an innovative solution for the roof trusses was to form them in reinforced concrete, a solution that provided the necessary fireproofing that was missing during the shelling of the cathedrals. Full records of the restorations, which eventually established the skills that currently aid many conservation sites, can be found at the Médiathèque de l'Architecture et du Patrimoine of the French Ministry of Culture (database Memoire).

A different solution was adapted for the new roof of the Great Hall in Stirling Castle in Scotland. The restoration of the 15 m span in 1999 aimed to fully reproduce the 1503 hammerbeam roof which was removed in 1800 when the castle was transformed into barracks and the Great Hall was partitioned. The reconstruction was based on a 1719 Ordnance Survey for the form and the contemporary Edinburgh Castle for the type of joints (Fawcett 2001; Lawrence *et al.* 1998).

Strathyre oak pieces were carefully sorted to form beams that were easy to assemble in order to compose the frame. The efficiency of the form and the amount of lateral thrust were assessed against the resistance of the walls using structural calculations. The original roof had caused the west wall to fail, so tie bars were placed between wall-head beams. The wall-head had to be levelled and a delicate process was followed to erect the roof gradually.

The result is a technical feat and the restoration of the rest of the Hall followed the same technical rigour. The need, however, for such a form is not easily justifiable after its absence for such a period or lack of reliable contemporary references. This was not the case in the church of the Eremitani in Padua: the polylobed timber roof was built by Giovanni degli Eremitani in 1306 and was destroyed by bombardment in 1944 (Fig. 6.34). Significant

6.34
The reconstructed polylobed timber roofs at the church of the Eremitani in Padua (courtesy Musei Civici di Padova)

Frauenkirche

Reconstruction from the ruins

The ruins from the 1945 bombardment stood as a silent monument to the disaster until Germany's Reunification (Fig. 6.35). As a monument to a historic turning point, the intention was to fully repristinate the original geometry, rather than creating a composition of its historic connotations, while making clear distinction between new and original fabric.

6.35
The ruins of Frauenkirche in 1945 (copyright SLUB/Deutsche Fotothek)

The conditions were to incorporate as much original fabric as possible, while following an archaeological reconstruction process that would reveal the original location of the pieces, the collapse dynamic and, also, key construction techniques. An equally critical parameter, however, was to correct the original design mistakes that precipitated the collapse of the dome and subsequently of the church.

A design procedure was developed to implement these considerations and several options were put forward (Wenzel 2007): change the geometry of the load-bearing structure (Zumpe); manipulate the stiffness (Leonhardt); use correcting forces (Wenzel-Engineering Partnership Frauenkirche); use of stronger materials (Siegel); or integration of additional structural elements (Wenzel).

Eventually the reconstruction process was based on a combination of the above, by pre-stressing the dome to cancel tensile forces using six flat steel ring anchors, assisted by strength classification of the recovered sandstone.

The project was clearly driven by national criteria and the technical decision to correct the original design error was almost inevitable (Fig. 6.36). The trauma of the ruined church has affected generations that were still alive when the decisions were taken. Germany certainly could have handled an interesting design solution as post-war reconstructions have shown (Altes Pinakothek in Munich or Gedächtniskirche in Berlin).

6.36
The reconstructed
Frauenkirche
Basilica (M.
Glendinning)

Maybe the project can be viewed as a reflection of Warsaw's experience. There is also an element of pride in employing high-quality engineering even in a project that essentially reproduced historic technology.

Appraisal of the solution

- Judging from the range of other post-war reconstructions (Warsaw, Berlin, Coventry), what other options would you consider?
- How should the reconstruction be planned if loads on the remaining ruins have to be controlled?

evidence and photographic records allowed the reconstruction of the roof, alongside the careful and technically innovative recomposition of fragments of frescoes by Mantegna (Ovetari Chapel).

In the case of partially damaged churches, rapid reconstruction seems a less controversial option. The Cámara Santa in Oviedo Cathedral is an important Christian and national monument that was destroyed in the first phase of the Spanish Civil War in October 1934. The reconstruction (1939–42) reused as much as possible of the original material (Fig. 1.4), which was carefully identified, and considered the attachment of the small building to the rest of the cathedral in order to minimise disruption (as published by the architect Luis Menéndez Pidal in 1960).

Similar national debates aiming to heal rapidly a recent bellic trauma guided the reconstruction of the Stari Most bridge in Mostar,

Bosnia-Herzegovina. The 1566 single-span pointed arch bridge was deliberately destroyed during the Yugoslavia War in 1993. Mostly new limestone was used in the 2004 reconstruction and the stability of the monument was improved by building concrete abutments (Armaly *et al.* 2004), in some contrast to the earlier faithful reconstruction of the Verona bridges after a similar bellic destruction (Cecchini 2004).

6.4.2 Restoring the fragment with new materials

The last case stands somewhere between the projects in the previous section and reconstructions that have not reused much of the original fabric. This is often the case in very large-scale reproductions where the fabric was not essential for the character of the monument, so as long as a sound base was provided, the aesthetics of the building could be reproduced by a faithful decorative scheme. In Warsaw most of the buildings of the Old Town (Stare and Nowe Miasto) as well as emblematic monuments (the Royal Castle, the Barbican, Wilanów Palace) were reconstructed following the deliberate destruction of 85 per cent of the city in 1944 (Lorenz 1964). The repristination followed soon after (mostly 1957–62) and was based on a pre-1939 architectural survey and the 1935 aerial photo-plan (Warszawa 1939). Most of the rubble from the ruins was reused and character was reinstated through recovery and integration of original elements or meticulous reproduction of the ornamentation, as in the ongoing project for the Royal Castle.

Similar approaches were followed in other damaged monuments, for example the Winter Palace in St Petersburg. Inevitably, perishable fabric such as timber or lime joints was substituted and it is impossible to avoid improvements or the introduction of current practices. The reconstruction of the Fenice theatre in Venice is a modern example of a complete reconstruction that updates the fabric and use. The 1792 theatre was destroyed by a fire in 1996, following reconstruction after another disaster in 1836. The project by Aldo Rossi (2003) re-created the theatre *com'era, dov'era*. New auxiliary spaces were carefully inserted within a new structure on flexible bases that avoided contact with the original fabric. The capacity was increased and improvements were made to the acoustic strategy (the ceiling profile was inclined following acoustic simulation), ventilation and, above all, fire protection and evacuation according to current practice (there is more on this aspect in section 6.6). Rossi also integrated as many of the original decorative fragments as possible, differentiating them slightly from the new supporting fabric.

The integration of fragments with new materials actually initiated the modern tendencies in conservation as highlighted in the work of G. Valadier (1762–1839). His seminal work on the partial reconstruction of the Arch of Titus at Rome's Forum (1819–22) was driven by his belief that 'more than the fabric, we need to conserve the architectural meaning (i.e. the form) of the Arch'. In this context, once he freed the monument from medieval additions, he reintegrated many of the original marble elements following meticulous archaeological studies. The missing portions of the entablature, which also stabilised the arch proper, were built in travertine, which permitted a subtle distinction from the original. However, this exemplary restoration may, in reality, have been driven by economic savings as the extensive need for new material might have been reduced by re-using most of the original fabric.

Further reactions to traumatic war destruction extend the range of the creative scale in the balance between the reuse of original material and repristinations of the original form. The reconstruction of the Alte Pinakothek in Munich (original design by Leo von Klenze, 1836) by Hans Döllgast (1946–57) is a key moment in the modern design approaches towards historic fabric. In contrast to the faithful reconstructions in Warsaw, Döllgast, representing contemporary German tendencies, chose to make evident the trauma of war and Germany's role, by emphasising the reconstructed portions, while not detracting from the neoclassical unity of the original. Plain

Neuesmuseum, Berlin

A tale of two fabrics

This project updates the cultural values of the Alte Pinakothek in a different historic context, seeking to heal the trauma of the ruined museum. The attempt to preserve the scars and integrate them into a coherent architectural composition presented a unique dichotomy where the remaining fabric of the neoclassical museum (1843–59) was restored by Julian Harrap and the ruined part was recomposed by David Chipperfield (2009).

Focusing on the latter section, the issue was to reinstate the museum and compose the new space with the fragments of the original and traces of the recent history. The technical challenges in creating an architectural and structural unity were addressed by using in-situ concrete as filler and for new load-bearing structure (Fig. 6.37).

6.37
The completed
south wing
of the
Neuesmuseum,
Berlin

A series of new concrete structures also created new galleries, such as the new Egyptian Hall and the main entrance (Fig. 6.38). The textures of concrete became crucial in balancing the effects and the precast elements used white cement and Saxonian marble aggregates (Bizley 2009). Recycled bricks from all over Europe were used and were thinly plastered to create a neutral tone for the roughened surfaces, both externally and internally.

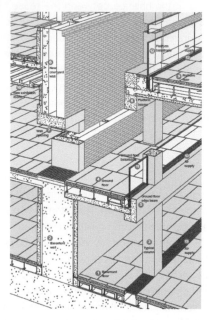

6.38
Detail of
the precast
concrete
structure
of the new
Egyptian Hall
(Bizley 2009)
– courtesy
Concrete
Centre and G.
Bizley

The project achieves its aims and despite the different 'language' between the two parts, they both attempt a composition of the fragments through the integrity of the essence of the original design.

Appraisal of the solution
- How would you choose the concrete mix and control its texture?
- What is the sequence of processes to reintegrate a fragmented wall using concrete or recycled bricks?

brick was used to reproduce the form of architectural elements but not their ornamentation, also in line with post-war austerity, very simply re-creating some of the monumentality of the interior at the same time.

The range of creativity used in the recomposition of a fragmented building, as was mentioned, might depend on the extent of disruption and the will to restore certain values of a monument. In the case of ancient monuments, it is usually confined to a small part, often with a didactic intention and materials are distinctive, as will be seen in the next section. Medieval castles offer more interesting opportunities for new structures that are sympathetic to the remaining textures. A creative recomposition is often triggered by the need to create enclosed spaces for exhibition or cultural activities. In the case of Kulturhaus Schloss Grosßenhain (2010) a new hall was created that combined with completion of a ruined tower, aiming to present the monument as a '*gestalteten Ruine*'. In the new Maritime Museum at Castillo de la Luz, Gran Canaria, the original forms were allowed to emerge once again by carefully adding new exhibition spaces in concrete in areas of the courtyard where similar volumes originally existed.

Koldinghus Castle, Denmark

Completion of a ruin with an independent structure

The castle has gone through a series of transformations until its partial destruction by fire in 1808. The restoration project by Inger & Johannes Exner followed after a very long period (1972–91) and was inspired by L. Kahn, establishing a dialogue between form (essence of architecture) and design (how form relates to period).

The ruin was appreciated for its own narrative and romantic aspects (Dirckinck-Holmfeld 1998). Reconstruction was dismissed by the restoration committee as there was not enough original material. The decision of the architects was to incorporate a completely new structure, going beyond the ruin or reconstruction debate (in their words 'We leave the ruins and we rebuild').

The choice of an impressive neo-Gothic structure in glulam timber allowed it to be placed completely detached from the ruins (Fig. 6.39). The elevation was completed in timber shingles as well, fastened on a frame that is suspended from the internal frame and simply touching the wall-head of the ruins (Fig. 6.40).

6.39
The new timber
structure at
the interior of
Koldinghus
(copyright Johannes
and Inger Exner)

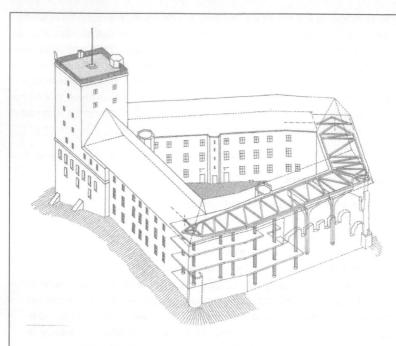

6.40
The overall
intervention
at Koldinghus
(copyright
Johannes and
Inger Exner)

Technically, the solution is fully reversible and the timber textures under the lighting arrangement create a new composition that updates and transforms the architectural and structural character of the monument.

Appraisal of the solution

- How would you accommodate the differential movement between the shingle facing and the rubble?
- Would you judge this highly creative solution as completely alien in the structural context of the original design?

6.4.3 Re-creating the essence of a form with new materials

The use of new materials in the Neuesmuseum and the attempt to 'freeze' the historic damage, transforming the building into the main exhibit makes one wonder what are the limits of how far new materials can be used to re-create the essence of the original design or the form of a most significant phase. In some cases, such as the new dome of the Reichstag and in most of F. Minissi's work, an evocation of the original form was sought, using modern materials, with an almost didactic intention. Other designers or building owners believe the re-creation of a building is achieved mainly through its use, relating mainly with its cultural context.

The new timber roof at the Oratory of the Forty Martyrs at Rome's Forum (Fig. 6.16) is a more subtle version of the completion in Koldinghus as the assumed original forms and volumes are reproduced in a design faithful to the essence of the original idea but using new materials such as timber. Such approaches can often have a strong didactic dimension and the modern design here has made the intervention attractive and informative, but not overwhelming. Didactic approaches can also be carried out on a much larger scale if lighter, almost temporary elements are used, as in the case of the Sanctuary of Apollo in Portonaccio.

Palazzo Carignano, Turin, Italy

New Baroque roof to maximise daylight

Refurbishment and restoration of Guarini's Baroque palace (1679–84) focused on the recovery of the Parlamento Subalpino underneath the elliptical pavilion façade tower, and the creation of an underground conference hall in the courtyard.

The objectives of the project by Andrea Bruno (1993) for restoration of the tower were to replace the roofs damaged by neglect and, above all, restore the idea of light from the original project of Guarini as a spine that vertically binds the space. Daylight through the drum of the tower has been obscured by a commemorative pediment erected in the memory of Vittorio Emanuele II in 1875 (Fig. 6.41).

6.41
External view
of Palazzo
Carignano, Turin

The main structural issue was the stability of the roof as a whole, since it was added later. Bruno decided to open up the area of the roof surrounding and blocking the windows of the drum (Fig. 6.42). This was in line with Guarini's original ideas for daylight as the architectural arrangement of openings in the pavilion has shown. Light becomes a constituent material of the project, similar to concrete and steel (Bosco 2000).

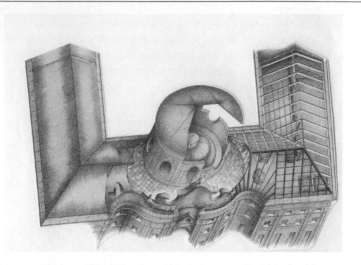

6.42
Axonometric view
of the intervention
around the central
lantern (copyright
Studio Bruno)

The shape of the area to be freed around the drum was determined by daylight studies. Glulam was chosen as it permitted trusses 27 m long with sufficient plasticity and strength to reinforce the edge of the unconventional geometry left by the circular intersection of the pitched roofs.

Appraisal of the solution

- What other alternatives can you think of for the re-creation of the plastic roof space?
- What strategy could be used to bring the daylight from the pavilion tower almost to the bottom of the building?

6.5 Correction of past design mistakes

This set of conservation problems does not result from negligence but ignorance and limitations inherent to the original design. As such they are often part of the historic and cultural values of the monument to be conserved and transmitted to the future. However, recovering the integrity of the original structural scheme does not ensure the building is also safe under either the previous conditions or the new ones the project dictates. In addition, it might contravene the building regulations or correct practice.

The quest for authenticity or the transmission of the historic dimension of the monument can actually compromise the very survival of its fabric. Therefore, the critical act that should characterise conservation is more relevant than ever in order to determine the extent to which some of the original errors must be reversed. The judgement of the structural engineers and, also, their understanding of the original scheme and the breadth of knowledge of techniques may produce solutions that are technically compatible with the current performance of the building. On many occasions, the current conditions rather than the original design should be considered as the datum for the present project.

The examination of recent interventions shows there is a range of approaches according to the severity of the problem that triggers the project. Some could be classified as part of cycles of maintenance, where the problem is resolved temporarily until further deterioration appears. This is the case of historic examples such as the dome of Hagia Sophia in Istanbul. The original design of the dome (AD 537) proved deficient and it partially collapsed in 558. The dome was completely rebuilt until a part was damaged at the earthquake of 989 and again in 1346. Repairs followed slowly after each event, until a major Ottoman strengthening in 1849 and the current conservation programme. Various hypotheses have been put forward for the original faulty design and it is accepted that the reconstructions improved the design by making the genetrix profile of the dome deeper or by strengthening the short piers at the base of the drum to function like ribs.

Many other monuments suffer from design limitations that affected their stability in the longer term. In the case of Gothic churches the dynamic equilibrium between the thrusts might not have been perfectly balanced initially, as the spectacular and ongoing deformations in Beauvais and Vitoria cathedrals show. Material degradation can increase the problem, especially when combined with vibrations from modern urban living, as occurred in the damage of many masonry towers in northern Italy, limiting the effect of conservative interventions, unless a drastic alteration to the loads is applied. Such limitations were inevitable because of ambition or unpredictable changes, but there is also an interesting category of designers actively ignoring any meaningful input from structural engineers, as happened with many of the early modernist architects. In general, such problems characterise those periods where rapid technological advancements are not assimilated at the same pace by the designers, either for cultural reasons or because of inability to disseminate the new practices.

6.5.1 Preserving poor design practice

Many iconic modernist buildings were inspired by the simplicity of industrial architecture and the potential of new materials like steel and concrete. The lack of systematic teaching in Architecture schools, combined with the creation of building codes only from the 1920s onwards, meant that many of the pioneers did not have a tested and safe knowledge of the systems they were trying to implement. This clearly supports the argument that architectural style and construction techniques do not always develop at the same rate (Ousterhout 2008) and is an indication of what happens when patrons, architects, engineers and contractors do not share the same amount of knowledge or appreciation for the efficiency of technical progress.

Some modernist buildings suffered mostly in terms of their performance rather than stability, so their restoration had to focus on these problems in order to return them to normal use and improve a tarnished image. This has often affected the restoration of housing tower blocks (Trellick Tower, London, or Park Hill Estate in Sheffield) or the work of Gillespie Kidd & Coia in Scotland, and especially Cardross Seminary (1965), because of the poor concrete detailing (Avanti Architects 2009).

Le Corbusier's Villa Savoie (1928–29) is probably the first genuinely conceptual building and his disregard for comfortable living, including a combination of wrong roofing details, caused the building to be gradually abandoned. Further negative modern legacies have been the flat roof (which has to be carefully detailed) and the pilotis system which was recently proved to play a controversial role in the high shear deformations that damage buildings during earthquakes. The restoration of Villa Savoie, promoted by A. Malraux (1958–67), attempted to balance the insufficient material properties of the fabric and cultural significance of its concept and forms (Murphy 2002).

Frank Lloyd Wright is another major designer who often disregarded engineering advice and did not propose technical solutions as innovative as his designs. Two of his emblematic buildings, Fallingwater and the Guggenheim Museum in New York (Meade and Hudson 2010), had to be restored in ways that would not compromise this historic limitation but essentially improve the original ill-conceived structural scheme (reinforcing the cantilevered terraces in Fallingwater or connecting properly the shear walls to the outer concrete ribbon in the Guggenheim).

Fallingwater

Preserve a fatal practice

F. L. Wright's seminal house in 1936 was characterised by the multiple cantilevered terraces and their smooth concrete finish. By 1941, however, extensive deflections and cracks became apparent, which were gradually attributed to inherent problems such as the daring spans and Wright's limited experience in RC structures.

As a consequence, the crucial slabs were incorrectly reinforced and the weight of wet concrete was disregarded. The building started deforming heavily and continuously (Fig. 6.43) and the window mullions eventually transferred loads from upper to lower terrace, further damaging the character of the interior. The deep cracks, in the natural forest conditions in which the house was situated, allowed corrosion of steel and spalled concrete.

Inspection with NDT was necessary to locate reinforcement and detailed mathematical (FE) analysis confirmed how incorrect detailing caused the problems (Silman and Matteo 2001).

6.43
Strengthening
of the heavily
deformed
cantilevers in
Fallingwater
(copyright
Western
Pennsylvania
Conservancy)

The decision was taken to arrest deflection with external post-tensioning so that the slab thickness was not altered, as the deformation had become part of the history of the house. PT strands run through blocks attached to sides of beams (Fig. 6.44), while the FE model allowed the degree the terraces would rise during tensioning to be properly controlled (Feldman 2005).

6.44
Detail of the
anchorage of
the tendons
in the
Fallingwater
cantilevers
(after Silman
and Matteo
2001)

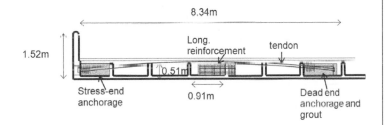

Appraisal of the solution
- What is the main challenge in post-tensioning already deformed slabs?
- How would you balance architectural unity with the historic values?

6.5.2 Reaching the limits of the fabric

Other categories of structures that start deforming soon after completion can be due to lack of knowledge (rather than disregard) of the limitations of their structural systems or the materials. The subtle difference from the previous category lies in the good intentions of the masons and the ambition in the design. Such problems of instability became more famously apparent in the Gothic period where the rapid increase in scale and slenderness was not supported by a contemporary dissemination of good practice. Failures were probably a more powerful means to communicate limits rather than performance and often this would represent a steep learning curve for the builders with beneficial effects even during phases of the same building. The design was the result of the building process, the need to manage local resources carefully and the slow setting times for lime mortar. Interventions to Gothic churches are therefore characterised by a direct link to the performance of the main structural elements, which highlights more clearly those values in the conservation of the buildings.

Two examples where erroneous design had catastrophic effects are Beauvais and Vitoria cathedrals. The partial collapse of the extremely slender piers in Beauvais on 29 November 1284 and the excessive deformations still active today (Taupin 1993) have traditionally defined the limits in the proportions of a Gothic church. Various hypotheses for the sources of the collapse have been put forward, such as the configuration of the offset junction (*porte-a-faux*) of the upper structure descending upon the spandrels of the aisle vaults or the inefficient propping action of the flying buttresses. The choir was restored in the following 40 years by inserting new supports between the piers, doubling the number of bays to six and transforming the vaults into sexpartite, re-using debris.

The delicate structure is still subject to accelerating deformations and vibrations today that have to be constantly monitored. It is almost impossible to return heavily deformed buildings to an original safe configuration without setting up a complex operation with many unknowns in the response of other parts or altering significantly the (apparently) failed scheme. Often it is a case of freezing or minimising the damage, as has been done in Vitoria Cathedral.

Vitoria Cathedral

Freezing the deformation

Vitoria Cathedral illustrates an interesting issue in Gothic architecture, i.e. historic remedies *after* significant lateral deformations have developed. The major construction phase occurred in the 16th century. Alarming deformations were already visible by 1647, followed by reconstruction of many vaults (Azkarate *et al.* 2001). Even this was not enough and a series of massive flying buttresses were built, together with the unique transverse arches in the nave at spandrel level of the aisle vaults.

The monument is clearly characterised by various additions of architectural forms and systems (more of an '*obra inacabada*', an unfinished fabric) and unfortunately attempts to re-create a stylistic unity in the 1960s by suppressing elements that did not 'fit' aggravated its conservation, both structurally and architecturally.

Since an emergency closure in 1994, a comprehensive restoration programme has been developed (Fig. 6.45). The movement of the church (specifically lateral spread) was carefully monitored in an effort to understand what was happening (Fig. 6.46). The heavily distorted configuration of the elevation (bulges of up to

30–40 cm can be observed) makes it ideal for the study of the behaviour of Gothic naves. The pattern is clearly caused by insufficient containment of the upper thrusts and eventually deformations of this scale were impossible to halt. Some of the ill-conceived 1960s buttresses were removed and many arches were rebuilt to their original curvature.

6.45
Restoring the deformed nave of Vitoria Cathedral – note the permanent arches bracing the transepts (copyright Fundación Catedral Santa María)

6.46
Vitoria Cathedral: comparison of measured deformations at the nave (Azkarate 2000) with those analytically assessed for Burgos Cathedral (B) (after Theodossopoulos 2004)

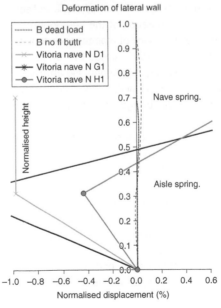

Deformation of lateral wall

· · · · · · · B dead load
- - - - B no fl buttr
—✕— Vitoria nave N D1
—✳— Vitoria nave N G1
—●— Vitoria nave N H1

Normalised height

Nave spring.

Aisle spring.

Normalised displacement (%)

Appraisal of the solution
- Considering the continuous evolution of the fabric, would it be more useful to create a new, functioning and correct configuration?
- The heavy deformations are clearly outside any reasonable or regulatory limits. Would you consider it safe and good practice to base the restoration on the present condition rather than a pristine original geometry?

The necessity to 'freeze' existing damage is often an opportunity to spectacularly preserve the history of a monument as an educational tool, especially for engineers, as the Stern Buttress at the Roman Coliseum or the deformed bays close to crossing towers in Selby Abbey in south Yorkshire, UK have shown.

6.5.3 Correction and updates

When, ultimately, errors lead to collapses, the decision to repristinate the original form has, inevitably, to address and correct them. If the analysis demonstrates that failure was triggered by an original fault, the critical act during design may choose to bypass the problem altogether. The crucial point is how much of the original fabric is reused. The decision to use extensive amounts, in the case of the Dresden Frauenkirche (Section 6.4) meant a solution sympathetic to the original configuration and function was chosen, including bracing of the base of the dome's drum.

Bell towers in Italy have often suffered spectacular and unpredicted failures. The 1585 Civic Tower of Pavia collapsed in 1989 causing one fatality. Gradual deterioration of the very inhomogeneous rubble core of the fabric combined with lack of maintenance resulted in a structure that could not cope with the vibration caused by traffic. There were no warning signs and the collapse was quite sudden. No attempts have been made to reconstruct the tower but, as with many recent emergencies, the failure has triggered conservative strengthening of many similar towers in northern Italy.

Similar origins of collapse and increased control measures in the aftermath had occurred much earlier, in the collapse of the Campanile of S. Marco in 1902, probably accelerated by the vibration of the bells and the insertion of a lift (Jokilehto 1999). A faithful reconstruction in reinforced concrete prevailed for compositional and historic reasons (1910), guided initially by L. Beltrami, prompting a significant international debate around the issue of authenticity in conservation, with opinions divided between full support or full opposition to the project.

In many historic cases, earlier parts of the structure are often substituted when the building advances during extensions, either because of perceived inadequate performance or simple stylistic demands. The original innovative vaults at the choir of Durham (completed in 1096), were probably a less confident version of the nave vaults, both belonging to the first period of the Cathedral (1093–1133). The choir vaults were replaced in 1235 during the addition of the Chapel of the Nine Altars at the east end, an action apparently triggered by the appearance of dangerous cracks, but, as often happened, it was probably an excuse to modernise the archaic vaults.

6.6 Update to current standards

Aspects of the modern built environment, such as accessibility, fire safety, thermal comfort and the associated legislation and regulations must also affect those interventions that aim to bring a monument back to public life. Often the demands are in direct contrast with the conservation conditions or character, and the examples discussed here show the creative approach of the designers or the progressive attitude of the authorities to reconcile vital discrepancies (Urquhart 2007). A crucial current problem is energy efficiency and compliance with updated regulations (such as the English Part L – see Wood and Oreszczyn 2004), with insulation or glazing being the usual areas of concern. Accessibility, fire safety and changes on imposed loads are three areas that have major effects on structural integrity and they are discussed in more detail in this section.

It is mainly buildings in public use, such as government agencies or company offices that have to be up to date with current regulations or practice and such interventions are often pressing and truly classified as refurbishment or retrofitting. It is the usual reason for upgrading a building as part of a maintenance programme, often occurring in regular cycles. In this context unfortunately many interesting solutions remain unknown or unpublished. Refurbishment can also be triggered by owners' liability, as non-compliance with regulations could lead to closure of the building or a high insurance premium. In addition, a building might have reached the limits of its service life or need to be upgraded for modern service systems, as happens with corporate estates (such as the refurbishment of HM Treasury in London). It is not about a simple re-wiring and floorplates may have to increase in section in order to accommodate new IT or HVAC systems.

Museum programmes are often the most imperative because of the sensitive contents and numbers of visitors, and they might demand structural or fabric modifications. The issues to address are conservation of the artefacts and comfort of visitors, and even environmental conditions demand interacting with fabric, as was seen in the case of the Palazzo Carignano discussed earlier. The design of new, semi-enclosed spaces of a larger scale that benefit from strategies that allow more daylight and natural ventilation was discussed earlier in the case of new atria. A smaller project is the Annenberg Court at the National Gallery in London, which resulted from the East Wing Project and the pedestrianisation of Trafalgar Square in 2004. Enclosing the newly opened courtyard from one side and using a translucent roof in ETFE provided the principal disabled access to the gallery and enhanced control of daylight in the spaces visitors were directed to (Ruffles 2008).

In other cases, the lack of accessibility or the limited available space to accommodate plant may be resolved altogether by creating a new space, often underground, as happened with the National Gallery of Scotland situated on The Mound, or the new entrance for the National Museum of Scotland, both in Edinburgh. The technical element is critical in the creation of such underground spaces and they will be discussed in Section 6.9 'New building and basements'.

Refurbishment due to higher demands for thermal comfort is more effectively achieved by the treatment of the envelope, either through enhanced insulation or the design of the façade as a system with environmental control strategies. In many modernist buildings or those where artistic rendering (plaster, frescos, colour scheme, etc.) decorates both sides of a wall, a complex intervention needs to be planned. The cases of the curtain wall in Boots D10, Nottingham or the Pirelli Tower, Milan, were triggered by strong degradation and will be dealt with in Section 6.7 'Securing the envelope'.

The examples that challenge the concept most are major refurbishments following catastrophic events or long-term neglect. Limitations of the original building become evident and it will have to be upgraded to make it suitable and safe for modern use. The Padiglione d'Arte Contemporaneo in Milan was reconstructed by J. Gardella, son of the original architect, after it was damaged by a bomb in 1993. As in La Fenice in Venice, the project took the opportunity to update some crucial details for the functioning of the building (comfort of visitors, conservation of exhibits, fire safety).

It is not often the case that the original designer will be invited to design the upgrade, prompting a fundamental question about the role of the creator in the service life of a building and whether it is acceptable that the original design process (if still represented faithfully) can still guide the later life of the building and its conservation. In many cases, the original architect even showed ignorance of what guarantees correct performance, as happened with Le Corbusier or F. L. Wright. It is usually the owners and policy-makers who make a clear choice not to re-employ the architect, for example Le Corbusier was never asked to restore his own buildings (see the debacle around Villa Savoie in Murphy 2002), while J. Utzon was only partially involved in the upgrades and modernisation of the Sydney Opera House (it was actually his son) and this was probably a political gesture linked to his dramatic exit from the project.

6.6.1 Accessibility
Ease of access for disabled visitors to, and users of, historic buildings is usually resolved by installing a lift or a series of ramps, both of which are, visually and structurally, quite intrusive.

Lifts require a rigid structure and, if built properly, they can be used to strategically strengthen the building. Insertion needs to be careful as they can be disruptive (a new lift caused the collapse of San Marco's Bell Tower in Venice). Wherever possible, the lift should be fully external, like the one at the Coliseum that was carefully inserted at the end of a gallery by P. Meogrossi (Orlandi 2011), or the glass towers at the Reina Sofia Art Museum in Madrid (by Ian Ritchie and Arup) which have become a feature of the gallery. In other cases, the lift has to be integrated into the fabric, often as a steel frame (e.g. Palazzo Altemps, Rome), and the associated technical problems will be discussed in Section 6.9. Finally, there is the curious case of lifts incorporated into domes that enable access mainly for tourists, as in St Peter's in Rome or the Berlin Dom.

Platforms in archaeological sites can be in sharp contrast or overwhelm the remains. Success of a subtle design depends on the careful location of supports and the choice of the materials, with steel and timber being the most suitable as they combine lightness with strength. The design has the potential even to enhance the experience by reproducing ancient paths or establishing viewpoints, as occurred in the ramps at the Fori Imperiali in Rome (Fig. 6.47) which re-create the medieval Via Biberatica which was destroyed during the major demolitions in the fascist period (Carbonara 2005).

6.6.2 Fire safety

The mechanisms with which fire affects historic building systems have been outlined in various sections in Chapter 3. The principles of fire safety engineering, as were summarised in Chapter 5, include a combination of active (fire suppression systems and fire alarms) and passive measures (fire and smoke barriers, space separation, fire prevention through minimising ignition sources, etc.).

Those retrofitting measures or post-fire design interventions that have a potential to alter the character of the building deal with material

6.47
Accessible viewing platforms re-creating the medieval Via Biberatica at the Fori Imperiali in Rome

protection or compartmentation. Exposed frames, especially modernist ones, may be deeply altered if painted with intumescent paints (wrought-iron or steel frames) or coated with a concrete screed, as in the case of concrete, brutalist architecture characterised by effects like board marks or material texture. In the latter case, if the cover of the reinforcement is damaged it has to be secured (see, for example, Park Hill Estate in Sheffield), and then active measures are more suitable, i.e. reduce the fire load and prevent fire from reaching the concrete surfaces.

Compartmentation, in principle, has to provide fire-resistance-rated walls and floors to form units that maintain a separating function during a fire. Some existing walls, such as partitions with some degree of cavity, could be improved but, in general, large gaps such as deep, through cracks in the masonry or voids at the joint of a roof with the wall, must be sealed so that smoke and, later, the flames cannot go through.

Progressive collapse in traditional fabric is often caused by the disintegration of vulnerable floors and roofs, which are expected to function as diaphragms, as happened in state houses like Penicuik House outside Edinburgh in 1899 or Lathallan House in 2006. Protection of the floor, therefore, can significantly improve the stability after a fire, but cannot be applied to exposed or decorated floors, unless from the outer side or between the floor joists.

It was also seen in Chapter 2 how concrete became more and more the material of choice for fire-proofing, promoting passive protection strategies. The fuel load is certainly reduced, compared to timber frames, and the reaction to the catastrophic fires in French cathedrals during the the First World War prompted the use of concrete, even in roof trusses.

Other passive strategies that might require a design intervention deal with the protection of escape routes, aiming to prevent the spread of smoke by creating a pressure differential (pressurisation or depressurisation) or inducing (opposed) air flow (CIBSE 2010). Air-handling systems, extraction, fire-resisting ductwork or natural ventilation strategies need to be introduced, which can be visually invasive or require changes in the pattern of use of the building.

Faculty of Engineering, University of Leicester

Lateral thinking

The iconic tower by James Stirling (1963) is characterised by the composition of the slender independent volumes that represent the functions of the building (hydraulic lab, lecture theatres, offices). The red tile envelope provides a playful contrast with the glazing of the main tower, and the whole volume is broken down by geometric devices, offsets and slender piers (Fig. 6.48).

6.48
The
Engineering
Faculty at the
University of
Leicester by J.
Stirling

Retrofitting for fire safety became necessary in 1996 as the stairs did not comply with current fire regulations, particularly for the provision of escape routes. It is Grade II listed, i.e. no external alterations are permitted, so no outer staircases or new openings could be added. The only solution available was to provide evacuation through existing stairs or use the lift shaft (Smith 1999).

The problem was resolved through pressurisation, where air is pumped to create a positive pressure at the protected escape routes (relative to the adjacent spaces) preventing smoke from reaching the area (Fig. 6.49).

6.49
Pressurisation
strategy around
the service shafts
at the Leicester
Engineering Faculty
(after CIBSE 2010
and Smith 1999)

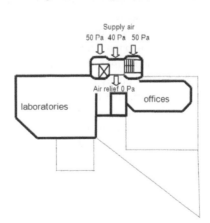

Essentially, the stairway was enclosed to make a pressurised shaft. The necessary large fans were sensitively moved to the tower roof, while air was pumped down through pipes incorporated in the enclosure.

Fire safety was completed with the integration of escape routes and exits (masking them with the red tile cladding, as with the original doors), and defining fire escape atria.

Appraisal of the solution
- How easily can this system be applied to other, traditional multi-storey buildings?
- What could be the limitations of natural ventilation strategies?

In the case of completely destroyed buildings, it is easier to rethink the whole fire safety strategy. In the new Fenice theatre in Venice, a building destroyed by fire, for example, a new RC structure was built and was designed for fire resistance and robustness up to the current standards. The stage was further compartmentalised with fire curtains and proper and safe evacuation strategies became possible.

6.6.3 Increased imposed loads
Building Regulations can also change the principles of the structural performance of a historic building by recommending that the fabric should be able to resist increased imposed loads (as a result of more demanding use) or adjusting the material and safety factors (often due to improved control in performance and material production). A typical case is existing masonry arch bridges: some of the techniques briefly described in Section 5.2 can affect the character as the traditionally considered high reserves cannot be easily quantified, and contractors would opt for major and conservative, yet invasive strengthening interventions.

Many historic masonry bridges also have listed status, which may conflict with the need to keep them functioning. The loads can be static or dynamic and can affect various areas of a bridge. Preliminary study of the capacity of bridges on a cross-country railway route for high-speed trains (McKibbins *et al.* 2006) showed that preventive strengthening would be required against the increased lateral load effects on the spandrels. Tie-bars anchored by pattress plates are always proposed as the best solution to clamp the spandrels, minimising traffic and visual disruption, altering however the role of the spandrels as containment.

Such a scheme was followed for the spandrel walls of the Egglestone Abbey bridge, when it became part of a route for HGV traffic in 1992. Stainless steel bars were fastened in a cementitious grouted sock anchor (McKibbins *et al.* 2006) and further works that avoided altering the appearance of the bridge concentrated on the voids (which were grouted so that the thrusts were contained uniformly) and the repair of some heavily weath-ered voussoirs (that were fixed with stainless steel pins). Furthermore, the arch barrel had to be tied to the repaired voussoirs so that continuity was restored.

A further range of imposed loads have sprung from the recent increase in scale of terrorist attacks, which can be classified as acciden-tal loads. The loss of life and the social and political impact of damage to usually public or high-profile property (Oklahoma City bombing 1995, WTC 2001, Manchester 1996) have led to a tightening of the rules: where such events were initially considered as accidental loads, nowadays very detailed practice and research has quantified the effect of firearm attack or explosion combined with complex fire patterns and evacuation strategies.

Queensberry House, Edinburgh

Blast protection

Regarding historic buildings, the conservation and conversion of Queensberry House as part of the new Scottish Parliament in Edinburgh in 2001–04 is indicative of the need for research and the sometimes controversial strategies triggered by external factors that have to be included in the design (Fig. 6.50). It became the finest townhouse in late 17th-century Scotland, after a major and masterful transformation by James Smith. The high quality of the fabric and internal fittings were compromised after three centuries of unsympathetic conversions and a full storey added in Napoleonic times (González-Longo 2008).

6.50
The restored exterior of Queensberry House, Edinburgh

The conversion by EMBT/RMJM (project architect C. González-Longo, 2001–04), aimed to conserve and expose the qualities of the original fabric. Although new floors had to be provided to withstand modern office loads, it was possible to conserve as a document the only two surviving original timber floors.

A blast consultancy was set up for the entire site, incorporating Queensberry House with the new buildings (Lewis *et al.* 2006). Straightforward design practice was improved with some 'enhanced measures' due to the Parliament's innovative cladding systems and materials, through explicit or experimental structural analysis. The transient pressures from detonation were modelled using US Army simulations and the aim was to provide an acceptable level of protection for the occupants.

As such, major efforts were concentrated on the performance of the windows and protection of the users against the hazards from glazing, especially because of the non-standard shape of the windows. 'Slam-shut' windows were developed that had a limited outward opening where the openable casement would be forced shut by the blast (Fig. 6.51). In Queensberry House a new timber casement was located inside the characteristic astragal windows and fastened on the new floors with a steel frame. The solution maintained the external character but inevitably blocked some daylight and the internal perception of the original fenestration.

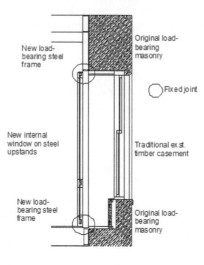

New load-bearing steel frame

Original load-bearing masonry

Fixed joint

New internal window on steel upstands

Traditional ex st. timber casement

New load-bearing steel frame

Original load-bearing masonry

6.51
New internal window frame as blast protection to occupants behind the existing timber window casement (Cristina González-Longo)

Appraisal of the solution
Can you think of similar requirements for other public buildings and what the implications could be if such methods are applied?

6.7 Securing the envelope

The gradual detachment of the envelope of buildings from their load-bearing fabric has characterised the development of structures mainly since the Gothic times and especially in modernist architecture and the creation of the curtain wall from the 1920s. Intervention to this area is conditioned by the

degree the envelope becomes a separate organism and almost a micro-struc-
ture in its own right. As such, stability and serviceability need to be considered
but very often the latter is more important and further reduction of deformation can result from improvements in the stability of the primary structure.

Very often historic envelopes concern mainly the ornamental
surface of a building, such as the non-load-bearing outer face of the wall in
ashlar in neoclassical Edinburgh. Restoration of this leaf improves also the
durability of the fabric and certainly the thermal performance of the inner
spaces. The bond and construction process, as seen above, does not create
a cavity wall, so repairs have to restore also the keying of the ashlar blocks
to the coursed rubble at the back.

Traceries in Gothic fenestration are often small structures on
their own, like the Dean's Eye, at the north transept in Lincoln Cathedral
(Fig. 6.52). Part of the Gothic phase of the building (1220), the window was
restored in 2006 as it was in poor condition (Clifton and Willis 2007). The
structure is made of stone bars tied with iron fixings. Various changes to
the original bars have taken place and, together with long-term deformations, this means that often the contained stained glass would provide a
diaphragm action and in-plane stability, which is not desired. Its behaviour
in various situations was analysed, and the consideration of wind load, in
particular, was complex, which demonstrated that there was a fault in the
original design and the Eye could not withstand the forces from the enclosing arches. The intervention, therefore, tried to improve the scheme and
especially the behaviour of the outer arches, ensuring that the bars could
transfer mainly compressive loads, or considering some elements as separate structures (like cantilevers) in order to control the amount of load they
could carry. Eventually the whole window was dismantled and reassembled
after all the stone was replaced with carefully selected limestone, incorporating hidden stainless steel reinforcement.

The design of Gothic structures attempted to disintegrate the
structure into a linear (almost frame) scheme in order to widen the openings

6.52
The restored Dean's
Eye in Lincoln
Cathedral (copyright
Geoff Clifton, Gifford)

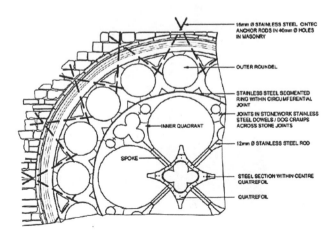

as much as possible. This ideal found full expression with concrete or steel frames in the 19th century and, eventually, independent cladding or glazing would create envelopes that define the character of the building and provide protection. In these cases, therefore, all design efforts should be concentrated on the envelope and the next examples show two approaches, with different degrees of subtlety, affected by the means available to the building owners.

The Pirelli Tower in Milan (Fig. 6.53) is an iconic design by Gió Ponti and P. L. Nervi (1956–60). Accidental impact from a light airplane in 2002 prompted a restoration in 2004 which was combined with the upgrade of the offices for the Lombardy Region (Salvo 2010). The project followed a rigour similar to pre-modern restoration projects in terms of preliminary research and historic-critical analysis and attempted to respect not just the geometric form ('*forma finita*') but also its materiality ('*consistenza materiale*'). The envelope is the key element and the solid parts were originally clad in Thermopan panels designed especially for Pirelli by St Gobain in 1959. The structure relied on the combination of the stair cores on the extremes, with the diaphragm action of the floors.

These floors had to be locally strengthened to allow them to keep working like diaphragms under the new loads prescribed by the current regulations. On the façade, the old glazing was replaced with laminated glass panels, 1900 × 1900 mm or 950 × 1900 mm, giving a total glazed surface of 20,000 m². The bottom-up curtain wall original construction facilitated the gradual and safe replacement, while the mullions were carefully reinstated.

Conservation of modernist architecture in general presents some very interesting challenges because of the closeness in time to the

6.53
Grattacielo Pirelli, Milan (Cristina González-Longo)

original project, especially its artistic and cultural values. This can cause a heavy deviation from the original character, as the similarity of the current version of the original technique may lead to shortcuts and loss of character.

The next example shows, alternatively, how contemporary technology can provide a range of commercially available materials to reproduce key elements that do not differ much from the damaged originals. Authenticity of the fabric is lost in both this and the previous example, but here it is the function that prevails, while the alterations that occur during the monument's life are due to degradation, as normally expected at the end of its useful life. The interventions can then be classified as maintenance that does not destroy any important historic layers.

Boots D10 Factory, Nottingham

Between maintenance and restoration

This seminal factory building, designed by Owen Williams in 1932 for the pharmaceutical firm Boots, is characterised by a genuine design of its glazing as a curtain wall which functions, in his own words, as a 'shell surrounding a process' (Fawcett *et al.* 1994).

The restoration by AMEC in 1991 focused on the refurbishment of the fenestration and minimised any change in the external steel patent of the glazing. It was decided to maintain the original mullion sightline widths of 20 mm and 47 mm and the original 2 : 1 : 1 mullion rhythm (Fig. 6.54). To reduce costs, off-site, quality-controlled fabrication from standard, readily available parts had to be chosen.

6.54
The curtain wall of Boots D10 factory in Beeston, Nottingham

At the same time, structural problems, such as concrete spalling, became evident, which affected the serviceability of the glazing. Frequent slab joints were also present in this long building, but eventually they had allowed water in at the interface of the external glazing with the concrete slab.

The original window system was cast-in at its base and a sliding joint was provided at the head for construction and thermal movement purposes (Fig. 6.55). Unfortunately this joint was later sealed and load was transferring through mullions, damaging the integrity of the windows. It was, therefore, important to restore the ability of the envelope to accommodate movement.

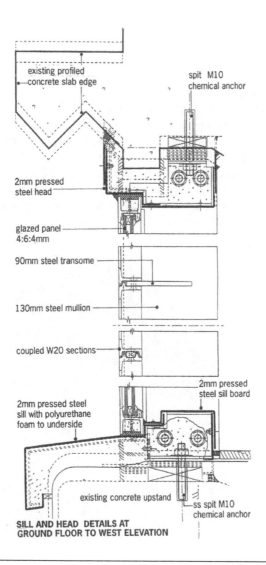

existing profiled
concrete slab edge

spit M10
chemical anchor

2mm pressed
steel head

glazed panel
4:6:4mm

90mm steel transome

130mm steel mullion

coupled W20 sections

2mm pressed
steel sill board

2mm pressed steel
sill with polyurethane
foam to underside

existing concrete upstand

ss spit M10
chemical anchor

**SILL AND HEAD DETAILS AT
GROUND FLOOR TO WEST ELEVATION**

6.55
Detail of the new
curtain wall in
Boots D10 (Fawcett
et al. 1994)

The subcontractor (Crittall) chose coupled W20 sections with 130 × 12 mm steel flats located in the original centres. The existing floor had variably deflected over the years and the slab edge varied by 50 mm at edge to 30 mm at head. In order to accommodate the tolerances, adjustable components were used and silicone pointing was minimised to a maximum of 20 mm.

Appraisal of the solution
- What do you think were the designers' choices and what had to be decided by the contractor?
- Both projects involved almost complete refurbishment. Would you propose a composition of new elements with authentic and intact ones?

6.8 Structural stiffness and stability

Stiffness and stability are two very valid problems that are strictly related with the responsibilities and art of the structural engineer and also constitute two crucial areas of structural theory. Most of the available repair techniques can improve successfully the stiffness of an individual element in order to make the distribution of loads more uniform and effective within a historic building being strengthened (Chapter 5). Similarly, tying a wall to its returns and floors provides effective bracing that eventually stabilises a larger portion of the building.

It is interesting, however, to examine the design perspectives of these two important engineering quantities when they affect a significant part of a building and, eventually, the techniques chosen for adjacent areas. Stability of an entire building is a major problem and the structure will have to be bonded throughout in order for the intervention to succeed. Increase of stiffness, on the other hand, might be part of the stability improvements as well as the strategies to redistribute the loads away from a delicate part or to enhance seismic resistance.

In historic masonry structures it is difficult to guarantee the continuity of the fabric that will ensure load redistribution as effectively as in concrete or steel structures. The concept, although commonplace in modern structures, has a rather theoretical potential in restoration, unless new fabric is inserted. The technology can be subtle but requires monitoring and various levels of control during planning and execution. Ultimately the result can completely alter the structural nature of the monument while causing little disruption to its aesthetic character.

6.8.1 Shear and bracing

In-plane loads are usually the main issue in stiffness improvements required in very flexible, wide-spanning interiors or where resistance to twisting forces (earthquake, settlement, irregular wind forces) has to be increased. Reinforcement (steel bars or fibres) might be suitable to improve tensile strength but the addition of more rigid materials in masonry, for example, can create compatibility problems, both mechanical and physical.

Bringing the entire section of an element into compression is one of the best passive ways to improve its in-plane resistance. This range of techniques introduces a new stress regime and alters the load distribution, affecting the authenticity of the original construction scheme ('*regola dell'arte*'). A quest for a return to a successful original performance might not be possible if that design failed or is inadequate for the new use of the monument. There is a subtle difference here from the concept of correcting past design mistakes in that the structure might simply be less adequate to withstand new recommended or unknown loads and the intervention can provide a better control over the performance.

Bracing or pre-stress can, however, be applied in a way that they are reversible. The increased compressive forces bond the parts of the structure better together and even seal some cracks that often are a cause of long-term degradation. It was seen in the re-creation of Frauenkirche how external bracing at the base of the dome corrected the faulty performance that led to the collapse and also relieved the rest of the structure from irregular loads.

A similar combination of correction of a faulty '*regola dell'arte*' and improvement of its structural scheme was applied in the reconstruction of the Cathedral of Noto in Sicily after the 1996 earthquake. The dome had already collapsed earlier in 1780 and 1848 and the masonry of the piers proved to be inadequate. Together with reconstructing them in a stronger bond, eventually the base of the rather tall drum of the dome was increased in thickness and braced externally with three steel rings, which were concealed by the cornice. The improvement of the bond of the dome is also expected to increase its resistance to seismic action.

Palazzo Altemps

Shear restoration

The medieval core of this building was extended and transformed in various stages in 1477, 1511 and 1568. More recently it has been converted into one of the three main archaeological museums in Rome (1984–93) to host ancient Roman statues restored to the eclectic tastes of Renaissance and Baroque patrons (collezione Ludovisi).

A major intervention was required when two load-bearing walls had to be removed to re-create the ballroom hall which hosts the Ludovisi 'Gaul Killing Himself' statue. It was found that these walls had been laid later and offset above the first floor theatre vaults, compromising their integrity.

The solution was to restore the shear strength of the missing walls (Figs 6.56 and 6.57). The upper part of the wall was kept and was formed into a stiff plate by being pre-stressed using stainless steel tendons and inserting a steel beam where it

6.56
Interior of main
hall in Palazzo
Altemps (with
kind permission of
the Ministero dei
Beni Attivitá
Culturali –
Soprintendenza
Speciale per i
Beni Archeologici
di Roma)

6.57
Detail of the
tendons in the
new lintels in
the main hall
(after Croci and
Scoppola 1993;
Carbonara 2005)

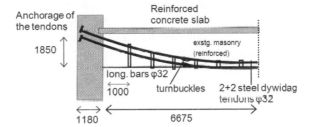

meets the joists (Croci and Scoppola 1993). The intervention was part of a wider rebonding of the building and strengthening of walls as the restoration project had many closed doorways re-opened.

Appraisal of the solution
- What would be the effect on the architectural integrity and structure of the Palazzo if a steel beam was used instead?
- Discuss the construction sequence of this operation.

6.8.2 Buildings' stiffness
Overall strategies of stiffness distribution and allocation to specific members, whether in the historic or new parts of an intervention, do not only improve resistance but can manipulate the stress regime, especially away from sensitive areas.

Heavily deformed Gothic cathedrals (Vitoria, Beauvais, Assisi) have profited to some extent from the potential of the concept. The strengthening and eventual stiffening of the vertical (piers) or even horizontal load-bearing elements (vaults, flying buttresses), often with some level of pre-stress, reduce the loads and deformations in other critical areas or shed them elsewhere. The effect of strengthening in critical areas became apparent for example in Beauvais when some historic iron ties between the slender flying buttresses were removed in the 1960s and were hastily reinstated in the 1990s after movement of up to 3 cm was recorded due to wind forces (Taupin 1993).

In the case of new structures in conjunction with historic fabric, the redistribution of stiffness can be better controlled as it can be evaluated accurately and executed using contemporary engineering methods. The concept has more applications in new structures inserted into existing ones, which is the subject of the next Section (6.9). In archaeological sites, concentrating stiffness in carefully selected locations can create transfer structures that shed loads away from sensitive or vulnerable areas.

The design of the new building by P. Zumthor for the Kolumba Diocesan Museum in Cologne (see Section 6.2) aimed to display the excavated ruins (Fig. 6.21) and combine gallery accommodation for the diocesan collections. The solution of a new concrete frame structure above the remains was carefully located within the ruins (Fig. 6.22), while stiff cores on the perimeter helped to create large transfer structures for the gallery and office floors above (Fig. 6.58).

In the area of seismic protection, the concept of robustness is critical and the use of light, innovative materials can prevent overloading the structure (Mazzolani and Mandara 2009). In the case of solid masonry buildings, reinforcement grids can be used to unify the response of the structure, improved even more by post-tensioning (see Fossanova Abbey). Alternatively, energy dissipation systems can be inserted into areas of highest concentration of stresses or where areas of high differences exist in stiffness or mass (and, consequently, in inertia and response to earthquake forces) that need decoupling.

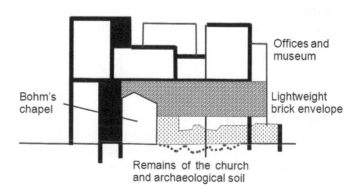

Offices and museum

Bohm's chapel

Lightweight brick envelope

Remains of the church and archaeological soil

6.58
Location of stiff cores in the Kolumba Museum, creating transfer structures over the ruins; transverse section (after Davey 2007)

6.8.3 Overall stability

Restoration and strengthening strategies are discussed in this section for technical problems that result from sources from the ground (geotechnics and earthquakes) or the building itself (unstable ruins or disturbed equilibrium). Settlement causes differential deformation on the structure and the strategy is either to strengthen the entire structure or control the settlement of the sections a building can potentially be divided into according to its stiffness or layout. Soil strengthening and piling are certainly the immediate solution to either stabilise the settlement or reverse it to a safe limit, but it is strengthening of the entire structure and its effect on the essence of the monument that is the focus of this section.

Slender bell towers are the most vulnerable structures under these circumstances and they can suffer heavy deformations, as seen in many Venetian bell towers (Santo Stefano, San Giorgio dei Greci, Frari). These collateral problems require interventions on the fabric like those at the Tower of Pisa during the latest works to reverse the settlement. Founded on weak, highly compressible soils, the gradual leaning became a cultural characteristic of the monument. After 1995 it became imperative to reverse the settlement to a safe limit and the preferred solution was to induce counter settlement by carefully extracting soil underneath the uplifting toe (Fig. 6.59). Combined with lead weights, 10 per cent of the inclination or 450 mm was recovered, moving the tower to a stable position, or at least its 17th-century position, reducing significantly the rate of further inclination (Croci 2000, 2001; Burland 2003).

The settlement had caused deformation of the cross-section of the tower, already worsened by the interruption from the staircase and seismic action, resulting in stress concentration at the first level. Preliminary, pre-stressed coated steel tendons were used to brace the two lower levels while the fabric was strengthened permanently by grouting, transverse tying of the tow faces of the masonry and radial tie bars for the smaller architectural elements.

6.59
Stabilisation of the Tower of Pisa by soil extraction (after Croci 2000; Burland 2003)

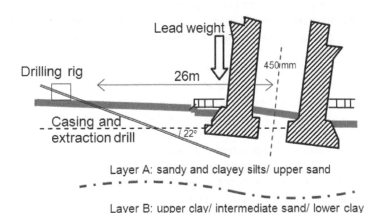

Layer A: sandy and clayey silts/ upper sand

Layer B: upper clay/ intermediate sand/ lower clay

Bell towers and Gothic spires can also suffer from vibrations due to bell ringing, wind loads or environmental actions. The spires of Burgos Cathedral had been strengthened earlier by bonding a wrought-iron truss to the tracery using cement. The differential expansion between the two structures had induced deformations and cracks in the bond and careful detachment and rebonding had to take place in 1995. In another case, the ringing of the bells in St Mary's Cathedral in Edinburgh had to be moderated, but the spires are under a regime of constant maintenance directed by the in-house workshop.

At the other end of the spectrum, large-scale buildings might require stability measures of commensurate scale. The interrupted portions of Gothic cathedrals (often the nave immediately after the transept) would be stabilised by arranging the last bays in a stepped fashion in order to contain the longitudinal thrust (as is still visible in Durham Cathedral today – see Theodossopoulos 2008). The transept of Beauvais Cathedral, after the final major failure due to the collapse of the extravagant spire in 1573, had to be stabilised in the long term as the church was never finished. Protection also had to be provided to the historic and venerable Basse-Oeuvre church standing at its foot. The solution was to infill the open arcades of the west elevation in 1605, creating a diaphragm tapering upwards like a massive buttress. Built in plain masonry, it offered the desired functional interruption in clear distinction from the ornate façades of the transepts. No design treatment, such as major tracery, was considered, both because of lack of funds and in a certain expectation that one day the cathedral would be completed. A similar treatment was given to the nave of Blyth Priory (Nottinghamshire) after its choir and transepts were demolished following the Dissolution in England.

Delicate structures, such as ruins, are even more vulnerable to stability problems. Castle Sinclair Girnigoe off Wick in Scotland has been battered by the elements since it was destroyed from 1680 onwards and its eroded tectonics make it a unique element of the landscape. The loss of contact between the eroded stone blocks, extended fragmentation and the location close to a cliff and sea spray made strengthening complex. Lime-based grout and mortar was used to re-bond the masonry and restore some continuity, as a material physically and mechanically compatible with the existing fabric.

Rigid block or drystone construction are, in theory, even more vulnerable in terms of stability. Many classical temples have been destroyed by seismic action or neglect, which manifested itself through loss of columns or vital continuity. The Parthenon in the Athenian Acropolis is an example of excellent resilience due to the quality of construction, the right proportions and the even distribution of stiffness across the monument (Ioannidou 2006). In the ongoing restoration project, often the best solution has been to restore the full contact and continuity between the members, i.e. return to the 'regola dell'arte' that proved successful as the monument has withstood many major earthquakes during its lifetime.

Parts of the monument occasionally had to be completed and the decision on the extent was also conditioned by contemporary good practice for seismic protection. Completion of some columns increases the vertical loads and consequently the friction between the blocks. The even distribution of rigidity can improve with the restoration of some walls and ceilings, as well as their connections. These principles were applied to the coffered ceilings of the Propylaia (Fig. 6.29) and the ceiling of the west pteron of the Parthenon.

Extending the discussion to the design effects of interventions to improve the earthquake resistance of masonry buildings, as was discussed in Chapter 3, there are various strategies that aim to restore the connections between the main elements, the diaphragm function of walls or floors and their integrity. In traditional or pre-industrial buildings there was a collective understanding of good practice and, often, failures occur when key elements have decayed or later additions (often in concrete) or removals have altered the structural scheme.

Therefore, restoring the seismic performance of key elements is often the best practice. Among the most compatible solutions in this sense is the reinforcement of wall junctions (especially the corners) or the improvement of the connections between the floors or the roof and the walls. The compressive strength of walls and piers or the flexural strength of floors can be improved with pre-stress (Karantoni and Fardis 1992). These techniques do not alter the function and character of the structural scheme, as long as the anchorage of the tendons does not demand disruption in the perimeter of the building.

Often a steel plate (preferably not a concrete edge beam) can be inserted to resist the pre-stress forces, as this does not cause severe visual disturbance. Aggressive solutions, however, such as shotcrete jackets, alter irreparably the function of the structure, while the extra thickness of the jacket damages the proportions of the building and destroys the original plaster, as has happened often in neoclassical houses in Greece. Sensitive analysis and preliminary critical study often leads to a better appreciation of the specific ways such buildings behave in earthquakes and, therefore, more respect for their original load-bearing function.

For more slender structures, the intervention might have to address the mechanical characteristics of the structure before any strengthening techniques can be applied. For example, the strengthening of the Annunziata campanile in Corfu, dating from 1394, focused on reducing significantly the modal periods of the masonry and improving its modulus of elasticity (Kouris and Karaveziroglou-Weber 2009). Overall it was suggested that the restoration of the diaphragm action of the floors should be effectively accompanied by improvement of the cohesion of the masonry and repair of the cracks.

Testing the proposed solutions with FE dynamic analysis according to Eurocode 8 showed that diaphragm action from post-tensioning had limited success but improvements in ductility and compressive strength could make a significant difference. Gunite jackets, however, are an inappropriate solution, while lime injections, although very compatible, must be validated for the high seismic forces. The rationale behind these proposals highlights a crucial problem with the lack of knowledge of the compressive strength of masonry, which often leads to over-conservative recommendations that prevent the exploration of solutions inspired by original construction techniques.

6.9 New building and basements

This category aims to discuss under the same conceptual umbrella the most extensive, almost aggressive alterations to historic buildings, where new fabric is inserted within either the superstructure or the substructure. In broad terms most of the interventions can be considered to impose their own structural scheme and even tectonics, with little attempt for a composition of the parts.

Most of the projects have resulted from commercial or usage pressures to recover as much space as possible from the property. Heritage protection, such as Grade B or Grade II listing in the UK, considers only the values of the exterior of a building and if enough pressure is applied the interior of a large property can be gutted, severing the structural integrity and unity between exterior and interior. There are often cases where the interior had a really poor distribution of internal spaces, unfit for office use, as has been said for the former General Post Office (now Waverley Gate) in Edinburgh. When value is challenged, however, the cultural aspects are irreparably damaged, even if they are defended by a pressure group or amenity society.

6.9.1 New building

The most straightforward case involves the construction of a new structure within a rather free interior, which is possible either because of its geometry or the state the building is in. There is more opportunity for a dialogue between the fabric and character of the existing and the new, as well as more choice of materials and systems to create interesting forms or materiality.

The conversion of Giles Gilbert Scott's grand Bankside Power Station (functioning from 1953 till 1982) into the Tate Modern art gallery in London by Herzog & de Meuron in 2000 kept the turbine hall as a cathedral-like exhibition volume (Fig. 6.60). Most of the new galleries were built within the Boiler House using a new, rather conventional, steel frame on a raft foundation (to avoid inserting piles). The technical challenges included the removal of the current work plant and maintaining stability of the new frame within the current envelope (Hirst *et al.* 2000).

A new gallery and viewing space was added at the top, which enhanced the powerful linear character of the station. Compatibility

6.60
Section through the
Tate Modern (Hirst
et al. 2000, copyright
Arup)

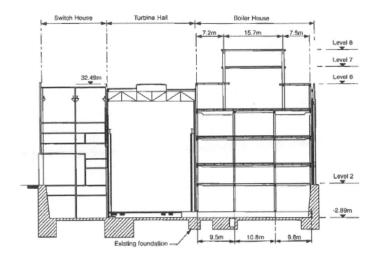

of the deformations of the new steel frame with the existing structure, and minimal internal visual intrusion of the former, were sought. Use and regulations demand high load capacity for new galleries and overall the new frame had a conventional bracing within four main cores. As the plan changes at Level 5, transfer beams were carefully added there. The construction process, therefore, required temporary bracing while the new structure was inserted and the existing removed, clearly a bottom-up process combined with expanding the capacity of the foundations. The intervention guaranteed a fruitful dialogue with the fabric even at a structural level, with the additions based on the same load-bearing principles and visual role.

Archivo Municipal, Valladolid

Superposition

The church of San Agustin (built between 1550 and 1627) gradually became a ruin after the Ecclesiastical Confiscations of 1835 (*Desamortización*). Its transformation into City Archives in 2002 developed through a process of superposition of the new archive and research facilities (Gallegos and González 2007).

Neglect caused the roof of the nave to collapse and because of its fragmented character the project 'selected' certain moments from the history and form of the church. The powerful image of the ruin was kept by consolidating the fabric, adding a neutral roof at the nave (Fig. 6.61) and inserting discretely the new structure into the aisles and lateral portions of the church, especially at the north side (Fig. 6.62). These areas gave the opportunity to recompose displaced elements and parts of the ancient cloister in a didactic way.

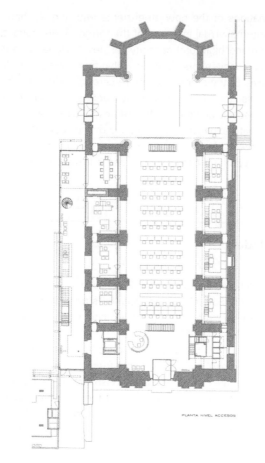

PLANTA NIVEL ACCESOS

Certainly the intervention managed to maintain the presence of the ruins that has consolidated in the public's memory. The structural scheme is rather neutral and almost reversible, while the assembly worked effectively around the ruin. The same can be said for the storage areas that were placed in the aisles on apparently separate steel frames that do not shed the heavy loads of the archive material onto the historic fabric. This quite subtle, almost ideal intervention became possible due to the conditions of the site and the scale of the Archives project.

Appraisal of the solution

- What do you think were the key factors that made it an intervention where the designers were in control rather than conservation specialists, even though a preliminary archaeological survey took place?
- What do you think are the main issues for archive storage and consultation in compliance with modern standards? Can you list the technical requirements?

It could be said that this project forms the basis for the discussion in this section about the insertion of new structures within a ruin or fragmented building. This discussion can be framed by two other cases, the timber structures at the Monastery of Fitero and the new floor structures at Queensberry House. The intervention in what was originally the library in the Monastery of Fitero attempted to recover the volumes of two main spaces with independent but rather massive structures in glulam timber (Alonso de Val and Hernandez Minguillón 2002). Over the kitchen a pyramidal roof was laid upon the existing masonry fixed on minimal support of a few steel pins to make clear the detachment and let daylight through the edge. A timber deck with 100 × 760 mm glulam timber beams replaced the missing vaults of the refectory and created a floor for the library.

The solid structures match well the tectonics of the bare ashlar walls in a clear structural detachment which, however, looks overwhelming in the refectory as no attempt was made to re-employ the proportions of the original stone vaults. In comparison with the more extensive and self-standing structures built in Koldinghus (Section 6.2), the integrity of the historic character and even original design of the spaces focused only in the recovery of the use of those areas.

At the other end of this range sit projects where new structures have been partially inserted. New composite floors had to be added in Queensberry House (Fig. 6.50) either to replace damaged timber floors or to couple them where they could still be integral but were incapable of carrying the new prescribed office loads (González-Longo 2008). What minimised the need for such new floors to provide vertical load-bearing and diaphragm action was the preservation of the masonry walls through repointing and lime grouting, restoring eventually the overall structural scheme of the monument.

6.9.2 Façade retentions

Although it seems there is a fine line compared with the previous category, very extensive removal and replacement of the entire internal structure are discussed in this section. The independence of the new structure is clearly pronounced: sometimes its performance depends on that of the preserved façade, but often this is used purely as an aesthetic or environmental envelope. Inserting a new building within a ruin, despite its technical challenges, is more straightforward conceptually as it removes the need to interact with any existing structure. The Visitor Centre at Whitby Abbey in Yorkshire by Stanton Williams (2002) integrated sensitively with the ruins of a mansion, aligning with the character of the original design but also offering a subtle preparation for the visit to the Abbey ruins. The new structure sits carefully adjacent to the main elevation or around the remaining brick walls, occasionally framing or offering viewing points with new platforms.

On the other hand, the new east end at Worksop Priory in Nottinghamshire offers an interesting architectural statement. The church was extensively reconstructed from 1845 following degradation after the Dissolution in England. The church was completed with a new east end designed by the architects Laurence King between 1970 and 1974, respecting hypotheses of the original configuration but in a contemporary style and materials (concrete). Functionally it is an integral part of the church, as well as its structural scheme, but technically it is a new building, which, however, did not necessitate altering any existing fabric. At the same time, it is the last in a series of new additions, but not in the same neo-Romanesque style the restoration of the choir had followed earlier in the 19th century.

Subtle interventions in this field affect only parts of the historic building with a clear detachment between the two fabrics. The Archaeological Museum of Asturias in Oviedo, Spain created a new building next to the preserved cloister of the San Vicente monastery, and tucked between an original façade and the deambulatory of the cathedral (Fig. 6.63). The project attempted to link the regular spaces of the cloister with the medieval urban fabric through a series of small atria that facilitated the exhibition and daylight conditions (Pardo Calvo and García Tapia 2004). The new concrete structure was then built almost independently and is visually detached from the original front façade.

A steel frame was used for the entirely new structure in a nearby project, the Museo de Bellas Artes (Mangado Beloqui 2010). Only the eclectic façade of the previous museum has been preserved here, which had to be kept stable during the works as all its original stability was lost and now depends entirely on the new frame. The temporary works and bracing of the weakened openings are typical of the heavy treatment of what is left of the historic fabric in this sort of project.

The complete obliteration of internal structure means the material character becomes incredibly fragmentary. The insertion process, on the other hand, however carefully it may be planned, will leave its mark on the remains. In even more demanding office and commercial uses, eventually

6.63
The Archaeological
Museum of Oviedo

6.63
The Archaeological
Museum of Oviedo

the new structure and services network have nothing to do with the original performance of the envelope.

Many of the technical issues are described in the extensive guide by CIRIA (Bussell *et al.* 2003; Highfield 1991) and were summarised in Section 5.2.5. The conversion of a Georgian building at 20–32 Baker Street, London into offices (Hansford 2002) offered some different solutions in the compatibility of a new steel frame with the brickwork envelope. To speed up the construction programme, the frame was built independently from the façade using a sacrificial dead shoring system which suspends the permanent beams from a series of adjustable stub columns. Then, in order to isolate the deformation of the long-span beams required for the office space from the masonry elevation, a pivoted connection within precast bearing blocks was fastened at the support of the beam. The contractors employed this method to keep the brickwork off the critical path and, to some extent, to permit some proper load-bearing function once the loads from the beams have transferred.

Most of such interventions, however, can be quite aggressive or overwhelming as the temporary works in the façade retention scheme for Marcol House in Regent Street, London shows (Fig. 6.64). Parking space can add further technical demands, involving large plant or delicate temporary works that add in cost, pushing for savings further down the construction programme. Maximising the office floor space is of paramount importance and thin beams can provide the acceptable room height, as in Baker Street. As an indication, the conversion of the GPO, Edinburgh, into Waverley Gate provided a total of 19,000 m² of office space in ten floors and basement car parking.

6.64
Façade retention
of Marcol House
in Regent Street,
London

The case of retained single elevations is a slightly different concept. The notion of the building that was enclosed by the façade is completely disregarded and the elevation becomes an architectural feature of the redevelopment. For example, the retained walls of the former council library became a part of the new façades of Wigan Life Centre, a local council regeneration scheme (Cooper 2010). They were propped up before the construction of the steel frame and the main new structure, and their stability was improved with some embedded concrete spines.

Attitudes that essentially ridicule any aspect of the existing building (technical, architectural, conceptual), and the concept of façade retention were mentioned in Section 5.2.5 (Lloyds of London, Marks & Spencer in Dublin). The Auditorio Paganini in Parma by R. Piano could fall into the same category as there was really little need to retain the external form of a historic yet modest sugar factory, situated outside any urban context or commercial need. The auditorium was created by gutting the building from all its floor plates, creating a shell with weak formal contribution to the design (Fig. 1.9).

6.9.3 Extending a structure and transferring loads
Commercial pressures can often push for more retail or office space in truly creative ways, in the case of multi-storey buildings upwards! This affects

modern buildings that can afford such extensions as straightforward continuations of their original frame in concrete or steel. This operation involves also a major overhaul of the envelope, a common case with most 1950s and 1960s office buildings whose systems have either exceeded their useful life (especially the ties of panels) or are obsolete in terms of occupants' comfort.

Four floors had to be added during the redevelopment of 80 Mosley Street in Manchester as rented office accommodation (Owen 2007). The project completely changed the bland appearance and layout of the building and gained more space through cantilevers, so the reinforcement of the existing concrete frame with bolted steel plates could be concealed. A new steel structure was fastened on top of the original concrete frame and some steel columns were inserted to reach all the way through the retained skeleton to the new foundation, reinforced with jack piling (see Fig. 1.7).

The almost purely technical problems of this project enabled a radical transformation of the building, where only the skeleton was retained, mainly for cost savings. In the case of buildings with historic character, such as the project to create more open retail and office space in 240 Regent Street, London, the effects of additions and removals on the character of the building must be considered. The 1920s Portland stone-clad building has a steel frame and, due to later design changes, two very delicate operations took place (Rowson 2008). A column had to be removed from the ground level to free access to the planned new lifts. A jacking operation was set up to remove the column and gradually transfer the load to new twin beams, permitting a slow and fully controlled transfer (Fig. 6.65).

Another operation, with stronger impact on the character of the conversion, was the alignment of the basement and ground columns along the façade with those of the upper floors. A new steel frame was carefully inserted in very confined conditions through holes in the slabs and the offset lower columns were removed once the upper floor loads were transferred onto the new frame through a mechanical Harjack system fixed on the top of its columns.

6.9.4 Basements

Basements are the other option in tight urban spaces. Such projects have been a necessity in crowded historic environments and were often created like vaulted cellars, during or after the construction of a building (see the case of undercrofts in Charlotte Square in Edinburgh or the entire substructure for the abutments of South Bridge, many of them still in use). In cases where the ground permitted it, basements would be dug directly into the ground, such as the city caves excavated underneath buildings in the rather soft bedrock of Nottingham (see the spectacular visualisations generated by 3D laser scanners for the Nottingham Caves Survey project).

From the 1960s onwards, major underground spaces were often created under churches to gain more space or link with existing crypts. The

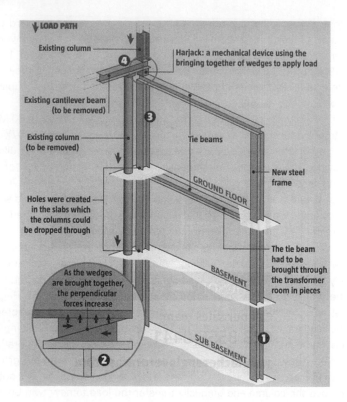

6.65
The jacking operation that enables load transfer to the new steel structure in 240 Regent Street, London (Rowson 2008)

undercroft and crypt of York Minster were created during the works to consolidate the foundations of the piers that support the crossing tower. The substantial operation to reveal and drill reinforcement at the pad foundations allowed the area in between to be archaeologically investigated, revealing many remains from the Romanesque cathedral (Feilden 2003). The area now combines with the crypt of the church and is fully accessible for educational purposes.

Other substantial operations are often set up to create underground space for the needs of the congregation. The intervention carried out by the J. W. Falkner and Sons building firm in 1975 at All Souls in Langham Place, London took advantage of the deep inverted-arches foundations that John Nash had built originally in 1823 in order to bypass the high water table (Meynell 2009). The project caused a small rise in the level of the church, which in addition was supported on rubber pads on top of the foundation walls in order to provide insulation from the vibrations of the Underground lines.

Other technical issues affecting design include waterproofing strategies such as tanking or extra insulation for thermal comfort. The sensitive designer can handle the often complex sequence of works and rely on the load-bearing capacity of existing undercrofts, following grouting with durable and reliable mixes, preferably based on lime. Heavy concrete jackets are often favoured in lining tunnels and other utility infrastructure, but should be avoided in historic settings (underground facilities in Galleria Borghese in Rome).

The stone barrel vaults in the basement of the Museum of Scotland in Edinburgh had been efficiently supporting the original concourse. A new access route was created in 2011, entering underneath both sides of the historic main entrance (Murphy 2011), making the existing undercroft into a new foyer for the museum at a slightly lower floor level. As a consequence, the walls supporting the vaults had to be eliminated and new doors had to open at the base of the building. Needle beams were used as temporary props to support the vaults and the wall while the lintel and its supports were created (Fig. 6.66). A lighter steel frame holds the valley of the barrel vaults and the new lintel at the gates is effectively hidden behind the facing stonework.

6.10 Concluding remarks

The discussion of the case studies in this chapter dealt with a wide range of interactions between new and existing fabric in the frame of a coherent architectural and structural design. In some cases, the complexity of the solutions went beyond the technical level as some problems required a stronger understanding of the conceptual issues, resulting in more sub-

6.66
The needle beams temporarily supporting the opening of a new door for the Museum of Scotland, Edinburgh

tle solutions such as the integration of new material to improve visual and structural integrity. This area, however, is essentially distinct from the more conservative approach to restoration that views interventions only as like-for-like repairs of damaged elements.

This book showcased historic buildings with interventions that recognised them as dynamic organisms, with aesthetic and cultural importance that has been meaningful continuously throughout history. As a consequence, the design of even subtle interventions requires comprehensive planning that treats with sensitivity the complex relations of the architectural, historic and cultural values, and a strong drive to integrate them with the current values these buildings have to serve. The discussions aimed to show how structural technology and design can also be a rich area in this exploration, not only in the case of those monuments where the structural identity prevails. It also aimed to show that modern analytical and technical skills, as well as structural forms (spatial structures, envelope systems), construction processes (concrete or prefabrication), spatial concepts, composition methods, etc. are very valuable in updating the historic built environment. Structural designers are key players in grasping and handling the necessary complexity towards successful and meaningful interventions for the progress of today's society through the understanding, integration and updating of historic architectural values.

Finally, it is also the hope of the book that the work of some designers and systematic advances in conservation theory reviewed here will become more widely available in the English language. Many of the buildings of such designers and theorists were presented here as key examples of a specific technique or approach. This was a deliberate abstraction from the complexity of the parameters that they had to address (in fact, many of these case studies were mentioned repeatedly throughout the discussion) and it was done in the educational context of this book. Communicating examples of such quality to young engineers and architects ultimately aims to engage them in a process of applying such quality not only to the restoration of monumental structures but also to the more humble examples of our historic architectural and urban environment, which are often of more direct relevance to our everyday life.

Projects gazetteer

For most of the projects discussed in this book, facts and figures have been collected in this gazetteer in order to avoid over-complicating the discussion of the intervention. The basic information concerns the original building (date, architect/engineer, use) and an outline of the intervention that is discussed in the book (architect and structural engineer, other consultants, the main or subcontractor) as well as some data such as the overall surface concerned, the approximate total cost and, a very important parameter, the client. It is inevitable that the information available is very variable, depending on the sources, so every effort was made to 'homogenise' the information and cover as many of these essential parameters as possible.

Acropolis monuments, Athens, Greece
Original buildings: Parthenon (Iktinos and Kallikrates, 447–438 BC), Erechtheion (421–406 BC), Propylaia (Mnesikles 437–432 BC), Athena Nike (Kallikrates, 427–424 BC)

Current project: Conservation and strengthening of the monuments on the Acropolis (works started 1985)

Design and construction agency: Υπηρεσία Συντήρησης Μνημείων Ακρόπολης (YSMA)

Client: Hellenic Ministry of Culture

Akrotiri, Santorini, Greece
Original: Minoan Bronze Age town destroyed by volcanic eruption in late 17th century BC

Current project: Bioclimatic protective enclosure to excavated site (2000–2011)

Architect: N. Findikakis

Main contractor: J&P-Αβαξ

Surface area: c.1.35 hectares

Client: Αρχαιολογική Εταιρεία (superintendent for Thera for the Hellenic Ministry of Culture: Chr. Doumas)

Cost: €30 million (2010)

All Souls at Langham Place, London, UK

Original design: John Nash, 1824

Current project: New undercroft space to include community hall, kitchen, dining room and toilets (1972–76)

Architect: Robert Potter

Main contractor: J. W. Falkner & Sons

Client: All Souls Langham

Alte Pinakothek, Munich, Germany

Original building: Alte Pinakothek (Leo von Klenze, opened 1836)

Current project: Restoration after severe damage during the Second World War (1946–57)

Architect: Hans Döllgast

Client: Bayerisches Landesamt für Denkmalpflege

Palazzo Altemps, Rome, Italy

Original building: Palazzo Riario-Solderini-Altemps (Martino Longhi il Vecchio, 1568)

Current project: Create spaces to house museum collections (Collezione Ludovisi), Archaeological Museum of Rome (1984–93)

Architect: Francesco Scoppola (Soprintendenza Archeologica di Roma)

Structural engineer: Prof. Giorgio Croci

Exhibition curator: Matilde De Angelis D'Ossat

Client: Soprintendenza Archeologica di Roma

Archaeological Museum of Asturias, Oviedo, Spain

Original building: San Vicente Benedictine monastery (1512–1791), restored in 1939 by Bobes and Menéndez Pidal and opened as Archaeological Museum in 1952.

Current project: Museo Arqueológico de Asturias, Oviedo (2002–08)

Architects: Fernando Pardo Calvo and Bernardo García Tapia

Structural engineers: NB35

Other consultants: GEASYT (services)

Main contractor: UTE MAO. Construcciones Alfredo Rodríguez S.A. and ARTEMON S.A.

Overall area: 5,665 m²

Client: Gerencia de Infraestructuras y Equipamientos de Cultura, Ministerio de Cultura

Cost: €6.2 million (2008)

City Archives, Valladolid, Spain

Original building: San Agustin Church and convent, 1550–1627

Current project: Rehabilitación de la Iglesia de San Agustín para Archivo Municipal de Valladolid, 2000–02

Architect: Primitivo González and Gabriel Gallegos

Structural engineers: Juan Carlos Alonso Monje

Other consultants: Foramen (Archaeology), Jorge Oteiza (artistic interventions)

Main contractor: NECSO

Overall area: 4,304 m² (useful area 3,090 m²)

Client: Ayuntamiento de Valladolid

Cost: €5.2 million (2002)

20–36 Baker Street, London, UK

Original building: Georgian residences

Current project: Conversion to offices and retail space

Architect: Norman & Dawbarn, with Erith & Terry (Quinlan Terry), 2003

Structural engineer: Whitby & Bird

Other consultants: English Heritage

Main contractor: Skanska UK Building Ltd; steel subcontractor: Bourne Steelwork

Overall area: 4,645 m² of commercial floor plate

Client: Loftus Family Property

Museo de Bellas Artes, Oviedo

Original buildings: Palacio de Velarde and Casa de Oviedo-Portal.

Current project: Reconstruction and extension of the museum along Calle de Rúa (1st phase 2009–12)

Architect: F. J. Mangado Beloqui

Structure and services: IDOM

Main contractor: SEDES

Overall area: 11,142 m²

Client: Consejería de Cultura y Turismo, Principado de Asturias

Cost: €16.6 million (budget in 2007)

Boots D10, Beeston, Nottingham, UK

Original building: D10 'Wets' Building, Boots Pure Drug Company (Owen Williams 1932)

Current project: Refurbishment of D10 Building (curtain wall, 1989–93)

Architect, structural engineer, quantity surveyor and main contractor: AMEC Design and Management

Overall glazed surface: c.3800 m²

Client: Boots Contract Manufacturing

Cost: £1,505/m² for the overall project; £364/m² for the envelope alone

Villa Romana del Casale, Piazza Armerina, Sicily, Italy

Original building: Late Roman Villa (part of a latifundia system). Excavations 1935–39 and 1950

Current project: Mosaics Museum, Villa Romana del Casale (1957–58)

Architect: Prof. Franco Minissi

Contributions: Cesare Brandi and Istituto Centrale del Restauro

Client: Direzione Generale di Antichità e Belle Arti, Ministero della Pubblica Istruzione

Castillo de la Luz, Las Palmas, Canarias, Spain

Original building: Castle of La Luz harbour (16th century), Las Palmas de Gran Canaria

Current project: Maritime Museum (2001–08)

Architect: Nieto Sobejano Arquitectos (Pedro Quero)

Structural engineer: NB35 SL, Jesús Jiménez Cañas

Other consultants: Aguilera Ingenieros S.A. (Pedro Aguilera): services; Proyectos Patrimoniales: archaeological survey

Main contractor: Dragados

Client: Ministerio de Fomento (Dirección General de Vivienda, Arquitectura y Urbanismo)

Crypta Balbi Museum, Rome, Italy

Original building: Part of portico built next to the theatre built by Lucius Cornelius Balbus in 13 BC in Campo Marzio.

Current project: Museo Nazionale Romano Crypta Balbi (1998–2000)

Architect: Maria Letizia Conforto and Massimo Lorenzetti (Studio F27: Massimo Lorenzetti, Stefano Castagna, Andrea Barazzutti, consultant Mauro De Luca)

Structure and services: Mario Bellini

Other consultants: Laura Vendittell (coordinator), D. Manacorda, M.S. Arena, P. Delogu, L. Paroli, M. Ricci, L. Saguì

Contractors: Cooperativa Archeologica (conservation), In.Co.Met. di L.Valdisolo & C. s.n.c. Teolo PD (steelwork), Mero Italiana s.r.l. (glass roof)

Overall area: c.3,000 m²

Client: Soprintendenza Archeologica di Roma, Ministero per i Beni e le Attività Culturali

Cost: €15.1 million (1998)

Roman Townhouse, Dorchester, UK

Original building: AD 4th-century town house close to the Roman town of Durnovaria

Current project: Shelter to conserve and present the mosaics of the Town House (1996–99)

Architect: Michael Howarth (John Stark and Crickmay Partnership)

Structural engineer: Paul Todd (AKS Ward)

Other consultants: Peter Cox (AC Archaeology Ltd), John Maine (artist), John Stewart (mosaics advice, English Heritage)

Contractors: D. J. Chutter Ltd (building), Robert A. Lovett (ironworker)

Client: Dorchester County Council

Einstein Tower, Potsdam, Germany

Original building: Einsteinturm, Erich Mendelsohn, 1920–22

Current project: Rehabilitation, 1997–99

Architect: Pitz & Hoh Werkstatt für Architektur und Denkmalpflege

Overall area: 685 m²

Client: Astrophysikalisches Institut Potsdam

Cost: €1.4 million

Fallingwater, USA

Original building: Fallingwater, country house designed for Edgar Kaufmann Sr by Frank Lloyd Wright (1936)

Current project: strengthening and repairs to Fallingwater (1995–2002)

Structural engineer: John Matteo (Robert Silman Associates)

Other consultants: Di Salvo Ericson Group (TDEG), post-tensioning consultant for the engineers

Contractor: VSL (post-tensioning) and Structural Preservation Systems, Inc (repairs)

Overall area: 495 m²

Client: Western Pennsylvania Conservancy

Cost: $11.5 million (total costs) (2001)

La Fenice, Venice, Italy

Original building: First project by Giannantonio Selva 1790–92, reconstructed after a fire in 1836 by Tammaso and Giambattista Meduna

Current project: Reconstruction of the theatre after the 1996 fire

Architect: Aldo Rossi

Structural engineer: Roberto Di Marco (steel and RC structures), Franco Laner (timber structures), Achille Balossi Restelli (special foundations)

Other consultants: Vanni Molinaro, TECSA S.p.A (security), Müller BBM, Munich (acoustics), AL Europe S.p.A. (mechanical services)

Project contractor: PH. Holzmann Bau Ag Süd, Romagnoli S.p.A., Al Europe S.p.A., Eleca S.p.A.

Overall volume: 160,000 m³

Client: Prefettura di Venezia

Cost: €53.1 million

Fishbourne Roman Palace, Chichester, West Sussex, UK

Original building: Roman Palace (villa and ancillary areas) for a local governor (c. AD 92)

Current project: Shelter for the protection and presentation of the mosaics (1968), following the site's discovery in 1960

Architect: Carden & Godfrey (Andrew Carden and Sheila Gibson)

Overall area: site is 150 m × 150 m

Client: Sussex Past (The Sussex Archaeological Society)

Fitero Monastery Library, Navarra, Spain

Original building: Cistercian Abbey of Fitero (16th century), abandoned due to the Ecclesiastical Confiscations of 1835 (*Desamortización*)

Current project: Reconstruction of the Library of the Monasterio de Fitero (1999–2001) after a collapse in 1997, and conversion to a museum

Architect: Miguel A. Alonso de Val, Rufino J. Hernández Minguillón (ah asociados)

Contractors: U.T.E. Fitero (Pavimentos de Navarra, S.A. and Arbeloa Construcciones, S.L.)

Overall area: 558.5 m²

Client: Ayuntamiento de Fitero

St Francis Basilica, Assisi, Italy

Original building: Basilica of St Francis of Assisi (Lower Basilica 1228–30; Upper Basilica 1239–53)

Current project: Reconstruction of vaults and roof and strengthening of the church following the devastating earthquake of 26 September 1997.

Architect: Prof. Paolo Rocchi

Structural engineer: Prof. Giorgio Croci (with Alberto Viskovic)

Other consultants: Giuseppe Basile (restoration of frescos – ICR), Costantino Centroni, Sandro Massa and material scientists from ICR and ENEA

Client: Sacro Convento di San Francesco d'Assisi and Commissione Pro Basilica di San Francesco (Ministero dei Beni Culturali)

Frauenkirche, Dresden, Germany

Original building: Frauenkirche, by George Bähr, Master Carpenter of the Dresden City Council (1726–43). Destroyed during the Allied bombardment of Dresden, 13 and 14 February 1945

Current project: Reconstruction of Frauenkirche, 1993–2005

Architect: Architekten- und Ingenieurgemeinschaft GmbH IPRO Dresden

Structural engineer: Ingenieurgemeinschaft Dr Wolfram Jäger/Prof. Fritz Wenzel

Other consultants: Prof. Jörg Peter (stress analysis engineer)

Main contractor: JV Frauenkirche Dresden (Walter Bau AG, Philipp Holzmann AG, Sächsische Sandsteinwerke GmbH, Germany)

Subcontractor: SPESA Spezialbau und Sanierung GmbH, Nordhausen/Thuringia, German, DSI GmbH, Germany (DYWIDAG Post-Tensioning Systems and GEWI)

Client: Stiftung Frauenkirche Dresden

Cost: €182.6 million (2005)

Domus, Fregellae, Italy

Original building: Domus and thermae of the Italic town of Fregellae (Frosinone), destroyed by the Romans after a revolt in 125 BC

Current project: archaeological park and shelter for three domus and thermae (1995–2008)

Architects: Laura Romagnoli and Andrea Marchi

Structural engineers: G. D'Angelo, A. Micozzi

Other consultants: Idrogeotec, Perugia (geophysics), Filippo Coarelli and University of Perugia (archaeology)

Client: Parco Archeologico di Fregellae, Consorzio Intercomunale Sistema Museale della 'Valle del Liri'

Great Court, British Museum, London, UK

Original building: British Museum built 1823–50 (architect: Sir Robert Smirke (1780–1867); Reading Room built 1852–57 by Sydney Smirke (1798–1877)

Current project: Create the enclosed Great Court and requalify the space; demolition 1998, roof construction summer 1999–Spring 2000

Architect: Norman Foster and Partners

Structural engineer (roof): Buro Happold (Steven Brown)

Contractor: Mace (construction manager, retail fitting; strong, existing supply chain relationships)

Other consultants: Emmer Pfenninger Partner AG (façade engineering); Buro Happold (services and M&E); Giles Quarme & Associates (historic building advisor)

Area of roof: Public courtyard = 6,700 m²

Cost: £30 million (1999) Roof: £100 million

Main spans: Plan 92 m × 73 m (6,716 m²); north span 23.8 m, east and west spans 14.4 m, south span 28.8 m

Total weight of roof: 420 t (or 75 kg/m² = 0.75 kN/m²), adding 60 kg/m² (0.6 kN/m²) from the glazing.

Hamburgmuseum (Museum für hamburgische Geschichte), Germany

Original building: 1914–22, Architect: Fritz Schumacher

Current project: Single-glazed gridshell over L-shaped courtyard (1989)

Architect: von Gerkan Marg und Partner, Hamburg

Structural engineer for the roof: Jörg Schlaich

Client: Museum für hamburgische Geschichte, Hamburg

Total plan area: c.900 m²

Total glazed surface: c.1000 m²

Major spans: 14 m on shorter arm and 18 m on the longer

Total weight of glazed roof: 50 t

Cost: DM3.5 million (1989)

Knossos Palace, Crete, Greece

Original building: Palace or local administrative centre in the Bronze Age Minoan Kingdom, destroyed gradually by earthquakes and fire around 1700 BC, 1450 BC and finally 1350 BC

Current project: Reconstructions by Sir Arthur Evans (1900–05)

Koldinghus Castle, Jutland, Denmark

Original building: Royal Castle of Koldinghus. Various building phases (1268, 1550, 1585, 1600, 1670, 1720) until gutted by fire in 1808

Current project: Restoration as a historic monument (1972–91)

Architect: Inger & Johannes Exner

Client: Palaces and Properties Agency, Ministry of Finance

Kolumba Museum, Cologne, Germany

Original building: Successive church buildings dedicated to Saint Kolumba occupied the site (I. 9th century AD; II. 11th centuryl III. 12/13th centuries; IV. c.1456) until destroyed in the Second World War (1945)

Current project: Kolumba Kunstmuseum des Erzbistums Köln (Diocesan Art Museum of Cologne) 2001–07

Architect: Peter Zumthor and Architekt BDB Wolfram Stein

Structural engineer: Ingenieurbüro Jürg Buchli, Haldenstein with Ingenieurbüro Dr Ottmar Schwab – Reiner Lemke, Köln

Other consultants: Ferdinand Stadlin Bautechnologie (services), Römisch-Germanisches-Museum, Köln (archaeology)

Contractor: Dyckerhoff AG, DyckerhoffWeiss (concrete), Unipor Ziegel (brickwork)

Overall area: 1,750 m² exhibition space

Client: Erzbistum Köln, Generalvikariat

Cost: €43.4 million (2007)

Kulturhaus Schloss Großenhain, Saxony, Germany

Original building: The medieval complex (never truly a Schloss) started in 1289 was reformed in 1663, destroyed in 1704 and rebuilt and extended from 1856 (functioning variously as a textile mill or gunpowder factory)

Current project: Kulturhaus Schloss Großenhain (2000–02)

Architect: Springer Architekten, Berlin

Structural engineer: Ingenieurbüro für baulichen Brandschutz Dipl. Ing. (FH) Winfried Bauer

Overall area: 16,000 m² (total), useful 8,000 m²

Client: Stadt Großenhain, Landesgartenschau Großenhain GmbH

Cost: €7.5 million (2002)

80 Mosley Street, Manchester, UK

Original building: Six-storey 1950s office block

Current project: Refurbishment, extension of floor plans with a new façade and upward expansion by four floors (2005–09)

Architect and structures: Robinson Consulting (Ralph Brade, Mohammad Rafiq, Brendan Harrison)

Other consultants: Spectrum MEP Consulting (building services)

Contractors: Hannans, LM Engineering (steelwork)

Overall area: After the extension each floor measures 25.5 m × 16.5 m (previous floor plan 23 m × 14 m). Total 2,787 m²

Client: Bilsdale Group

Cost: £5.5 million (2007)

Museum of Scotland, Edinburgh, UK

Original building: Royal Museum of Scotland, Captain Francis Fowke (1861)

Current project: Royal Museum Redevelopment (2003–11)

Architect: Gareth Hoskins Architects

Structural engineer: David Narro Associates

Main contractor: Balfour Beatty Construction Limited

Other consultants: Max Fordham (building services engineers), Turner & Townsend Management Solutions, Ralph Appelbaum (exhibition designer)

Client: National Museums of Scotland

Cost: £46 million (2011)

Neptune Court, Greenwich London, UK

Original building: Greenwich Hospital. West central wing by Daniel Asher Alexander (1806); south and west ranges by Philip Hardwick (1862)

Current project: National Maritime Museum (1995–99)

Architect: Building Design Partnership (BDP), Rick Mather Architects

Structural/services and lighting engineer: Building Design Partnership (BDP)

Contractors: Eiffel Construction Metallique (roof), Jasper Jacob Associates (exhibition)

Overall area: 7,700 m²; Court plan: 45 m × 54 m

Cost: £19.9 million (1999)

Neues Museum, Berlin, Germany

Original building: Neues Museum, Friedrich August Stüler (1841–59), heavily damaged in the Second World War (1945)

Current project: Reconstruction and restoration (2003–09) to host the Egyptian Museum, the Papyrus Collection and the Museum of Prehistory and Early History

Architects: David Chipperfield (south wing and overall control) and Julian Harrap (north wing and conservation), with Lubic & Woehrlin

Structural engineer: Ingenieurgruppe Bauen

Contractors: Ernst & Young Real Estate GmbH (management), Dressler (precast concrete subcontractor, Egyptian Court), Pro Denkmal GmbH (conservation)

Other consultants: Kardorff Ingenieure Lichtplanung (lighting), Michele De Lucchi (exhibition design), Ingenieurbüro Axel C. Rahn GmbH (services)

Overall area: 19,600 m²

Client: Staatliche Museen zu Berlin, Stiftung Preussischer Kulturbesitz

Cost: €200 million (2011)

Cathedral of Noto, Sicily, Italy

Original building: Cattedrale di San Nicoló di Noto (1694–1703)

Current project: Reconstruction of the damaged cathedral after the earthquake of 1996

Architect: Salvatore Tringali

Structural engineer: Roberto De Benedictis

Other consultants: Prof. C. Gavarini, Prof. L. Binda, Prof. G. Baronio, Prof. S. Tobriner, Prof. M. Maugeri

Contractors: Donati Spa, La Cattedrale S. C. a.r.l.

Client: Diocesi di Noto, Prefettura di Siracusa

Cost: €2 million (1999–2007)

Olite Castle, Navarra, Spain

Original building: Castillo de Olite, Palacio de los Reyes de Navarra (13th and 14th centuries); destroyed during the War of Independence (1813)

Current project: Archaeological restoration (1937–63) and Parador (1965)

Architecture and structures: José and Javier Yárnoz Larrosa

Contractor: Oficina Técnica del Palacio real de Olite (José Yárnoz Larrosa)

Client: Diputación Foral de Navarra (Institución Príncipe de Viana, Sección de Patrimonio Artístico)

La Olmeda Roman Villa, Pedrosa de la Vega, Palencia, Spain

Original building: Rural Roman villa of 4th century AD. Discovered in 1968

Current project: Museum and visitors centre, shelter to protect and present mosaics (2004–09)

Architect: Paredes Pedrosa Arquitectos

Structural engineers: Alfonso G.Gaite GOCAITE sl, UTE La Olmeda

Other consultants: Nieves Plaza (services)

Overall area: Total 7,130 m², Oecus (mosaics) 175 m²

Client: Diputación de Palencia

Cost: €6.3 million

Orangery, Chiddingstone Castle, Kent, UK

Original building: 18th-century Orangery, Chiddingstone Castle

Current project: Restoration of Orangery for public events and new gridshell roof (2007)

Architect: Peter Hulbert Architects

Structural engineer: Buro Happold (Richard Harris)

Contractor: Carpenter Oak & Woodland Limited (alternating chestnut lathe gridshell), IPIG (glass)

Client: Chiddingstone Castle Trustees

Oratory of the Forty Martyrs, Forum, Rome, Italy

Original building: Excubitorium or Giuturna Nympheum (Roman Imperial period), then probably Oratorio dei XL Martiri (destroyed by earthquake in AD 847)

Current project: Presentation of the Oratory and protection of the frescos (2000)

Architect: Claudia del Monti (Soprintendenza Archeologica di Roma)

Other consultants: Ernesto Monaco, Dieter Mertens

Contractor: E. A. L., Carlo Usai (restorer)

Overall area: 10.50 m × 8.50 m, assumed height 40 Roman feet (*c.*11.84 m)

Client: Soprintendenza Archeologica di Roma

Palacio Cibeles, Madrid, Spain

Original building: Palacio de Comunicaciones (GPO)

Designers: Palacios Ramilo y Joaquín Otamendi Machimbarrena (1904–18)

Current project: Central chambers for the City of Madrid

Architect: Arquimatica SLP (Francisco Rodriguez Partearroyo, Angel Martínez Díaz, Francisco Martínez Díaz, David Márquez Latorre)

Structural engineer: Schlaich Bergermann und Partner (roof)

Contractor: Lanik I, SA (roof)

Overall area: Roof (Galería de Cristal) 3,000 m²

Client: Ayuntamiento de Madrid

Cost: €100 million (2011), for the roof and transformation to a cultural centre

Palazzo Carignano, Turin, Italy

Original building: Palazzo Carignano (Guarino Guarini 1679–83), Parlamento Subalpino (Carlo Sada, 1848), Italian Parliament (Gaetano Ferri and Giuseppe Bollati, 1863), Museo Nazionale del Risorgimento (1937)

Current project: Restoration of Baroque apartments and underground facilities in courtyard (1982–85)

Architect: Andrea Bruno

Structural engineers: Adriano Strutturini and Gianni Vercelli

Contractor: Guerrini

Overall area: 5,000 m²

Client: Museo Nazionale del Risorgimento

Pirelli Tower, Milan, Italy

Original building: Grattacielo Pirelli, by Gió Ponti and P. L. Nervi (1956–61)

Current project: Offices for regional council of Lombardy (2004), following an airplane accident in 2003. Restoration of façade

Architect: Adriano Crotti (façade)

Structural engineer: Antonio Migliacci, Maurizio Acito

Contractors: A.T.I./ISA, Grassi & Crespi, Marcora, BMS Progetti, St Gobain (cladding panels)

Other consultants: Renato Sarno Group (façade consultant), Prof. Giorgio Torraca.

Overall area: 127 m height (31 floors), total surface 24,000 m², glazed surface *c.*20,000 m²

Client: Regione Lombardia

Tower of Pisa, Italy

Original building: Torre Campanaria, Santa Maria Assunta Cathedral, Pisa (1173–1254)

Current project: Stabilisation of the tower to 5 degrees (or 3.99 m of displacement), 1990–2001

Engineering committee: Comitato internazionale per la Salvaguardia della Torre di Pisa: M. Jamiolkowski (Chairman), J. B. Burland, R. Calzona, M. Cordaro, G. Creazza, G. Croci, M. D'Elia, R. Di Stefano, Jean de Barthelemy, G. Macchi, L. Sampaolesi, S. Settis, F. Veniale, C. Viggiani

Client: Ministero dei Beni Culturali, Ministero dei Lavori Publici

Cost: 14 billion lira (2000)

Queensberry House, Edinburgh, UK

Original building: Town house for the 2nd Duke of Queensberry (James Smith, after 1695), built on a previous building

Current project: Offices for the Scottish Parliament, Edinburgh (2001–04)

Architect: EMBT/RMJM (Cristina González-Longo)

Structural engineer: Ove Arup & Partners (Edinburgh)

Contractor: Bovis Lend Lease

Overall area: *c.*2,000 m²

Client: Scottish Parliament

Cost: £11 million (2002)

240 Regent Street, London, UK

Original building: Dickins & Jones Department Store, designed by Sir Henry Tanner (1919, 1921 and 1938), 224–244 Regent Street, London

Current project: Mixed-use development in 240 Regent Street (2009)

Architect: Rolfe Judd

Structural engineer: Cundall Johnson & Partners (Gary Rollison)

Contractor: Bovis Lend Lease

Steel contractor: Bourne Steel

Overall area: Retail 10,500 m²; office 8,000 m²; restaurant 1,200 m²

Client: Shearer Property Group & Delancey

Courts at Richelieu Wing, The Louvre, Paris, France

Original building: 1854–59 (Pavillon Richelieu: 1852–57), by Visconti and Lefuel

Current project: Unification of galleries and access to create the Grand Louvre (1988–93)

Architects: I. M. Pei, Leonard Jacobson, Yann Weymouth, Stephen Rustow, C.C. Pei

Associate architects: Agence Macary, Agence Wilmotte, Agence Nicot

Other consultants: R.F.R. – Peter Rice, Michael Francis, Ian Ritchie (roofs, skylights, galleries); Serete S.A., Paris (systems); Claude R. Engle, Washington, DC, Observatorie, Paris (lighting)

Client: Etablissement Public du Grand Louvre

Cost: FF880 million (1993)

Great Hall, Stirling Castle, UK

Original building: Great Hall of the Royal Palace in Stirling (1502), transformed into barracks in the 18th century

Current project: Re-create hammerbeam roof in Scottish oak (1990–99) and ongoing restoration

Architecture and structural engineering: Historic Scotland

Contractor: Carpenter Oak & Woodland (oak roof)

Other consultants: Richard Fawcett (Historic Scotland)

Overall area: 423 m²

Client: Historic Scotland

Cost: £8 million (2000)

Tate Modern Gallery, London, UK

Original building: Bankside Power Station, Giles Gilbert Scott, 1953–82

Current project: Conversion to Tate Modern Gallery, 2000

Architect: Herzog and de Meuron with Sheppard Robson

Contractor: Schal (Construction Manager)

Structural, civil, building services, fire, acoustic engineer: Ove Arup & Partners (John Hirst, Faith Wainwright, Jim Shaw)

Quantity surveyor: Davis Langdon & Everest

Overall area: Galleries 7,827 m², turbine hall 3,268 m²

Client: Tate Gallery Projects Ltd

Cost: £134 million

Trinity Apse, Edinburgh, UK

Original building: Church of Trinity College and Trinity Hospital (1460); demolished in 1848 to make way for Waverley Station

Current project: Stones from original church were numbered and stored in the open till 1872, when just the apse of the church was rebuilt in Jeffrey Street, Edinburgh (1877)

Architect: John Lessels

Veranes Roman Villa, Asturias, Spain

Original building: Villa Romana de Veranes (4th century AD)

Current project: Museum of the residential villa and shelter for the mosaics (2007)

Architect: Manuel García García

Other consultants: Carmen Fernández Ochoa, Fernando Gil Sendino, Ricardo Mar

Overall area: c.90 m²

Client: Ayuntamiento de Gijon

Cost: €730,438 (2005), design and construction of the shelter

Wigan Life Centre, UK

Original building: City Library

Current project: Wigan Life Centre (2011)

Architect: AStudio (Max Rengifo) and LCE Archimed

Structural and services engineer: Morgan Sindall

Contractors: Morgan Sindall, Elland Steel (steel truss roof)

Overall area: 5,358 m² (Information and Learning Zone)

Client: Wigan Metropolitan Borough Council, Wigan Leisure and Culture Trust

Cost: £50.5 million (entire regeneration scheme)

Bibliography

Architectural conservation

Bellini, A. 1995. Il tempo del restaurare, il tempo del conservare. *Recto-verso. Itinerari nei luoghi delle architetture e delle estetiche*, Vol. 1, pp. 3–10, Milan: Guerini studio.

Bluestone, D. 1999. Academics in tennis shoes: historic preservation and the Academy. *Journal of the Society of Architectural Historians*. 58(3) Architectural History 1999/2000 (Sept.): 300–307.

Boito, C. 1883. *Prima Carta del Restauro*. A presentation at the III Conference of Architects and Civil Engineers of Rome.

Bonelli, R. 1963. Il restauro architettonico, entry Restauro. In *Enciclopedia Universale dell'Arte*, vol. XI, Florence: Sansoni.

Brandi, C. 1963. *Teoria del Restauro*. Turin: Einaudi.

Carbonara, G. 1976. *La reintegrazione dell'imagine: problemi di restauro dei monumenti*. Rome: Bulzoni.

Croci, G. 1987. La metodologia scientifica nello studio dei monumenti: il complesso del Tabularium e Palazzo Senatorio in Campidoglio ed il Colosseo. In Sandro Benedetti and Gaetano Miarelli Mariani (eds) *Saggi in onore di Guglielmo de Angelis d'Ossat*. Rome: Multigrafica Ed. (Quaderni dell'Istituto di Storia dell'Architettura; N.S. 1/10.1983/87), pp. 533–542.

ICOMOS 1993. Guidelines for education and training in the conservation of monuments, ensembles and sites. Available at: http://cif.icomos.org/pdf_docs/Documents%20on%20line/GUIDELINES%20FOR%20EDUCATION%20AND%20TRAINING%20IN%20THE%20CONSERVATION.pdf (accessed February 2012).

Jokilehto, J. 1999. *A History of Architectural Conservation*. Oxford: Butterworth-Heinemann.

Lourenço, P. B. 2002. Computations of historical masonry constructions. *Progress in Structural Engineering and Materials*, 4(3): 301–319.

Matero, F. G. 2007. Loss, compensation, and authenticity: the contribution of Cesare Brandi to architectural conservation in America. *Future Anterior*, 4(1): 45–58.

Miarelli Mariani, G. 2002. Durata, intervallo . . . restauro: singolarità in architettura. *Quaderni dell'Istituto di storia dell'architettura*, N.S. 34–39 (1999–2002), pp. 33–46.

Roca, P., Cervera, M., Gariup, G. and Pelá, L. 2010. Structural analysis of masonry historical constructions: classical and advanced approaches. *Archives of Computational Methods in Engineering*, 17(3): 299–325.

Riegl, A. 1903. *Der moderne Denkmalkultus, sein Wesen, seine Entstehung*. Vienna: W. Braumüller.

Stanley Price, N., Kirby Talley, M., Melucco Vaccaro, A. 1996. *Historical and Philosophical Issues in the Conservation of Cultural Heritage*. Los Angeles: Getty Conservation Institute.

Stubbs, J. H., Makaš, E. G. and Bouchenaki, M. 2011. *Architectural Conservation in Europe and the Americas*. Hoboken, NJ: John Wiley & Sons.

Technical aspects (structures, construction, environment)

Acland, J. H. 1972. *Medieval Structure: The Gothic Vault*. Toronto: University of Toronto Press.

Adams, H. 1906. *Building Construction* (reprint). Donhead St Mary: Donhead Publishing, 2011.

Alvarez de Buergo Ballester, M. and Gonzalez Limon, T. 1994. *Restauración de edificios monumentales*. Madrid: CEDEX.

Andrews, D. 2003. *Measured and Drawn: Techniques and Practice for the Metric Survey of Historic Buildings*. Swindon: English Heritage.

Architects Registration Board (ARB). 2003. Criteria for the Prescription of Qualifications. Available at: www.arb.org.uk/templates/includes/qualifications/criteria/Prescription%20of%20Qualifications%20ARB%20Criteria.pdf (accessed September 2011).

Ashurst, J. 1988. *Practical Building Conservation: English Heritage Technical Handbooks. Vols 1 to 5*. Aldershot: Gower Technical.

Atamturktur, S. and Laman, J. A. 2010. Finite element model correlation and calibration of historic masonry monuments: review. *The Structural Design of Tall and Special Buildings*. New York: John Wiley & Sons, Ltd.

Augenti, N. and Parisi, F. 2009. Experimental data analysis on mechanical parameters of tuff masonry. *Workshop on Design for Rehabilitation of Masonry Structures – Wondermasonry 2009*, Lacco Ameno, Italy, 8–10 October.

Barberot, E. 1895. *Traité des constructions civiles*. Paris: Baudry & Cie Editeurs.

Bechmann, R. 1981. *Les racines des cathedrales: l'architecture gothique, expression des conditions du milieu*. Paris: Payot.

Beckmann, P. and Bowles, J. 2004. *Structural Aspects of Building Conservation*. Oxford: Elsevier Butterworth-Heinemann.

Bennett, D. 2001. *Exploring Concrete Architecture Tone, Texture, Form*. Basel: Birkhauser.

Benvenuto, E. 1990. *An Introduction to the History of Structural Mechanics. Vol. 2: Vaulted Structures and Elastic Systems*. New York: Springer-Verlag.

Berto, L., Saetta, A., Scotta, R. and Vitaliani, R. 2002. An orthotropic damage model for masonry structures. *International Journal for Numerical Methods in Engineering*, 55(2): 127–157.

Burn, R. S. 1873. *Building Construction: Showing the Employment of Brickwork and Masonry in the Practical Construction of Buildings*. London and Glasgow: W. Collins.

Bussell, M., Lazarus, D. and Ross, P. 2003. *Retention of Masonry Façades: Best Practice Guide*. CIRIA C579 document. London: CIRIA.

Calderini, C. 2008. Use of RC in preservation of historic buildings: conceptions and misconceptions in the early 20th century. *International Journal of Architectural Heritage*, 2(1): 25–59.

Campin, F. 1883. *A Treatise on the Application of Iron to Bridges, Girders and Roofs* (3rd edition). London: Crosby, Lockwood & Co.

Bibliography

Carbonara, G. 1990. *Restauro dei monumenti. Guida agli elaborati grafici*. Naples: Liguori Editore.

Carbonara, G. 2010. Foreword. In C. F. Carocci and C. Tocci (eds) *Antonino Giuffrè. Leggendo il libro delle antiche architetture: aspetti statici del restauro: saggi 1985–1997*. Rome: Gangemi.

Castigliano, A. 1879. *Théorie de l'équilibre des systémes élastiques et ses applications*. Turin: A.F. Negro.

Charleson, A. 2005. *Structures as Architecture*. Oxford: Elsevier.

Chartered Institution of Building Services Engineers (CIBSE) 2010. *CIBSE Guide E: Fire Safety Engineering*. London: CIBSE.

CIRIA (Construction Industry Research and Information Association) 1994. *Structural Renovation of Traditional Buildings*. Report 111. London: CIRIA.

Clemente, R., Roca, P. and Cervera, M. 2006. Damage model with crack localization – application to historical buildings. In P. B. Lourenço, P. Roca, C. Modena and S. Agrawal (eds) *Proceedings of Conference 'Structural Analysis of Historical Constructions'*. New Delhi: Macmillan India Ltd, pp. 1125–1135.

Clifton-Taylor, A. 1972. *The Pattern of English Building*. London: Faber and Faber Ltd.

Concrete Society 2004. *Design Guidance for Strengthening Concrete Structures Using Fibre Composite Materials* (2nd Edition). Report of a Concrete Society Committee.

Croci, G. 1998. *The Conservation and Structural Restoration of Architectural Heritage*. Southampton: Computational Mechanics Publications.

Croci, G. 2001. *Conservazione e restauro strutturale dei beni architettonici*. Turin: UTET Universitá.

Davey, A., Heath, B., Hodges, D., Ketchin, M. and Milne, R. 1981. *The Care and Conservation of Georgian Houses*. Edinburgh New Town Conservation Committee, London: The Architectural Press.

Dernie, D. 2003. *New Stone Architecture*. London: Laurence King.

Dietz, K. and Schürmann, A. 2006. Foundation improvement of historic buildings by micro piles, Museum Island, Berlin and St. Kolumba, Cologne. In *7th ISM International Workshop, Schrobenhausen*.

Dimes, F. G. and Ashurst, J. 1998. *Conservation of Building and Decorative Stone*. Oxford: Butterworth-Heinemann.

Donghi, D. 1906. *Manuale dell'architetto*. 10 volumes (1906–1925). Turin: UTET.

Docci, M. and Maestri, D. 2009. *Manuale di rilevamento architettonico e urbano*. Rome: Laterza.

Esposito, D. 1998. *Tecniche costruttive murarie medievali: murature 'a tufelli' in area romana*. Rome: 'L'Erma' di Bretschneider.

Feilden, B. M. 2003. *Conservation of Historic Buildings* (3rd Edition). Amsterdam and London: Architectural Press.

Fitchen, J. 1961. *The Construction of Gothic Cathedrals: A Study of Medieval Vault Erection*. Oxford: Clarendon Press.

Forsyth, M. 2007a. *Materials and Skills for Historic Building Conservation*. Volume 3. Oxford: Blackwell Publishing.

Forsyth, M. 2007b. *Structures & Construction in Historic Building Conservation*. Oxford: Blackwell Publishing Ltd.

Fouchal, F., Lebon, F., Titeux, I. 2009. Contribution to the modelling of interfaces in masonry construction. *Construction and Building Materials*, 23: 2428–2441.

Ganz, H. R. and Thürlimann, B. 1982. Tests on the biaxial strength of masonry (in

German). Report No. 7502-3, Institute of Structural Engineering, ETH Zurich (after Lourenço 2002).

Gauthey, M. 1843. *Traité de la construction des ponts*. Publié par M. Navier, Tome premier. Liége: Chez Leduc.

Giovanetti, F. 1998. Typological process towards urban rehabilitation: The Manuale del Recupero of Rome. In Attilio Petruccioli (ed.) *Typological Process and Design Theory*. Cambridge, MA: Aga Khan Program for Islamic Architecture.

Giuliani, C. F. 1992. *L'edilizia nell'antichitá*. Rome: La Nuova Italia Scientifica.

Glendinning, M. and Muthesius, S. 1994. *Tower Block: Modern Public Housing in England, Scotland, Wales and Northern Ireland*. London and New Haven: published for the Paul Mellon Centre for Studies in British Art by Yale University Press.

Hendry, A. W., Sinha, B. P. and Davies, S. R. 1997. *Design of Masonry Structures*. Spon Press: London.

Henry, A. 2006. *Stone Conservation. Principles and Practice*. Donhead St Mary: Donhead Publishing.

Heyman, J. 1966. The stone skeleton. *International Journal of Solids and Structures*, 2: 249–279.

Heyman, J. 1977. *Equilibrium of Shell Structures*. Oxford: Clarendon Press.

Heyman, J. 1995. *The Stone Skeleton. Structural Engineering of Masonry Architecture*. Cambridge: Cambridge University Press.

Highfield, D. 1991. *The Construction of New Buildings behind Historic Facades*. London: Spon.

Hollis, M. 2005. *Surveying Buildings*. London: RICS.

HS 2001. *TAN 22: Fire Risk Management in Heritage Buildings*. Edinburgh: Historic Scotland.

HS 2005. *TAN 28: Fire Safety Management in Heritage Buildings*. Edinburgh: Historic Scotland.

HS Technical Conservation Research and Education Group 2007. *COST Action C17: Built Heritage: Fire Loss to Historic Building Final Report* Edinburgh: Historic Scotland.

Huerta, S. 2006. Geometry and equilibrium: the Gothic theory of structural design. *The Structural Engineer*, 84(2): 23–28.

Hume, I. 1997. The structural engineer in conservation. *The Structural Engineer*, 75(3): 33–37.

IAEA 2002. *Guidebook on Non-Destructive Testing of Concrete Structures*. Vienna: IAEA.

Institution of Civil Engineers (ICE) 1989. *Conservation of Engineering Structures*. London: Thomas Telford.

IStructE (Institution of Structural Engineers) 2008. *Guide to Surveys and Inspections of Buildings and Associated Structures*. London: Institution of Structural Engineers.

Jagfeld, M. 2000. *Tragverhalten und statische Berechnung gemauerter Gewölbe bei großen Auflagerverschiebungen*. Schriftenreihe des Lehrstuhls für Tragwerksplanung Prof. R. Barthel. Munich: Shaker.

Joint Board of Moderators (JBM) 2009. Guidelines for accredited MEng degree programmes leading to chartered engineer. Available at: www.jbm.org.uk/uploads/JBM111_MEng.pdf (accessed December 2011).

Karantoni, F. 2004. *Masonry Construction. Design and Repairs*. Athens: Papasotiriou. (Originally published in Greek: Καραντώνη, Φ. 2004. Κατασκευές από Τοιχοποιία. Σχεδιασμός και Επισκευές. Αθήνα: Εκδ. Παπασωτηρίου.

Karantoni, F. and Fardis, M. 1992. Effectiveness of seismic strengthening techniques for masonry buildings. *Journal of Structural Engineering, ASCE*, 118(7): 1884–1902.

Kooharian, A. (1953). Limit analysis of voussoir (segmental) and concrete arches. *Journal of the American Concrete Institute ACI*, 89(1): 317–328.

Lourenço, P. B. 2005. From fracture mechanics to case studies: the issue of cultural heritage. In J. Galvéz *et al.* (eds) *Anales de Mecánica de la Fractura*. Almagro, pp. 10–17.

Lourenço, P., Rots, J. G. and Blaauwendraad, J. 1998. Continuum model for masonry: parameter estimation and validation. *Journal of Structural Engineering, ASCE*, 124(6): 642–652.

Lorenz, W. 2009. Foreword. *Third International Congress on Construction History*, Cottbus, 20–24 May. Chair of Construction History and Structural Preservation of the Brandenburg University of Technology.

Lupp, J. R., Fitz, S., Baer, N. S. and Livingston, R. A. 1998. *Conservation of Historic Brick Structures: Case Studies and Reports of Research*. Donhead St Mary: Donhead Publishing.

Mainstone, R. J. 1998. *Developments in Structural Form*. London: Architectural Press.

Mallouchou-Tufano, F. 1998. *The Restoration of Ancient Monuments in Greece, 1834–1939*. Athens: Library of the Archaeological Society in Athens. (Originally published in Greek: Μαλλουχου-Tufano, Φ. 1998. Η αναστήλωση των αρχαίων μνημείων στη νεώτερη Ελλάδα (1834–1939). Βιβλιοθήκη της εν Αθήναις Αρχαιολογικής Εταιρείας αρ. 176.

Mazzolani, F. M. and Mandara, A. 2009. Seismic protection of historical buildings: the PROHITECH project. *4th International Congress 'Science and Technology for the Safeguard of Cultural Heritage of the Mediterranean Basin'*, Cairo, Egypt, 6–8 December.

McCann, D. M. and Forde, M. C. 2001. Review of NDT methods in the assessment of concrete and masonry structures. *NDT & E International*, 34(2): 71–84.

McKibbins, L. D., Melbourne, C., Sawar, N. and Sicilia Gallard, C. 2006. *Masonry Arch Bridges: Condition Appraisal and Remedial Treatment*. CIRIA Report C656, London: CIRIA.

McMillan, A., Gillanders, R. and Fairhurst, J. A. 1999. *Building Stones of Edinburgh*. Edinburgh: Edinburgh Geological Society.

Mediatheque de l'Architecture et du Patrimoine. 2011. Base Memoire: sites et monuments. Available at www.mediatheque-patrimoine.culture.gouv.fr/ (accessed November 2011).

Milani, G., Lourenço, P. B. 2010. A simplified homogenized limit analysis model for randomly assembled blocks out-of-plane loaded. *Computers & Structures*, 88(11–12): 690–717.

Mundell, C., McCombie, P., Bailey, C., Heath, A. and Walker, P. 2010. Limit-equilibrium assessment of drystone retaining structures. *Geotechnical Engineering, Institution of Civil Engineers*, 162(4): 3–12.

Nacente, F. 1890. *El constructor moderno, Tratado Teórico y Práctico de Arquitectura y Albañilería*. Barcelona: Ignacio Monrós y Compañía.

Newlands, J. 1900. *The Carpenter and Joiner's Assistant*. London: Blackie & Son.

Nicholson, P. 1823. *The New Practical Builder and Workman's Companion* (1st Edition). London: Thomas Kelly.

Nicholson, P. 1828. *A Popular and Practical Treatise on Masonry and Stone-cutting* (1st Edition). London: Thomas Hurst, Edward Chance & Co.

Ochsendorf, J. 2006. The masonry arch on spreading supports. *The Structural Engineer*, 84(2): 29–35.

Ousterhout, R. 1999. *Master Builders of Byzantium*. Princeton, NJ: Princeton University Press.

Ousterhout, R. 2008. Architecture and cultural identity in the Eastern Mediterranean. *DFG Schwerpunktprogramm 1173, Integration und Desintegration der Kulturen im europäischen Mittelalter*.

Purchase, W. R. 1929. *Practical Masonry: A Guide to the Art of Stone Cutting Comprising the Construction, Setting-Out, and Working of Stairs, Circular Work, Arches, Niches, Domes, Pendentives, Vaults, Tracery Windows, Etc*. New York: London Stone Crosby Lockwood.

Robson, P. 1999. *Structural Repair of Traditional Buildings*. Shaftesbury: Donhead Publishing.

Ross, P. 2002. *Appraisal and Repair of Timber Structures*. London: Thomas Telford.

Rovira y Rabassa, A. 1897. *Estereotomía de la piedra*. Barcelona: Librería y estampería artistic.

Ruddock, T. 2008. *Arch Bridges and their Builders 1735–1835*. Cambridge: Cambridge University Press.

Ruffles, A. 2008. *Lighting Strategies for New Space Roofs on Historic Buildings*. Dissertation for the degree of BEng Structural Engineering with Architecture, University of Edinburgh.

Sanabria, S. L. 1982. The mechaniazation of design in the 16th century: the structural formulae of Rodrigo Gil de Hontañón, *Journal of the Society of Architectural Historians*, XLI(4): 281–293.

Sinha, B., Ng, C. L. and Pedreschi, R. 1997. Failure criterion and behavior of brickwork in biaxial bending. *Journal of Materials in Civil Engineering ASCE*, 9(2): 70–75.

Skempton, A. W. 2002. *A Biographical Dictionary of Civil Engineers in Great Britain and Ireland: 1500–1830*. London: Thomas Telford.

Sorour, M. M., Shrive, N. G., Parsekian, G. A., Duchesne, D., Paquette, J., Mufti, A., and Jaeger, L. 2009. Experimental determination of E and G values for stone masonry walls. In F. M. Mazzolani (ed.), *Protection of Historical Buildings, PRO-HITECH 09*. London: Taylor & Francis Group.

Sutherland, R. J. M. 2000. A century of innovation: structural engineering 1900–2000. *The Structural Engineer*, 78(7): 4 January 2000.

Sutherland, R. J. M., Humm, D. and Chrimes, M. 2001. *Historic Concrete: The Background to Appraisal*. London: Thomas Telford.

Swailes, T. 1998. *Structural Appraisal of Iron-Framed Textile Mills*. London: T. Telford.

Ulm, F. J. and Piau, J. M. 1993. Fall of a temple: theory of contact applied to masonry joints. *Journal of Structural Engineering ASCE*, 119(3): 687–697.

Ungewitter, G. G. and Mohrmann, K. 1890. *Lehrbuch der gotischen Konstruktionen*. Leipzig: Tauchnitz (after Heyman 1995).

Urquhart, D. 2007. *Conversion of Traditional Buildings. Application of the Scottish Building Standards*. HS Practitioners Guide 6, Edinburgh: Historic Scotland.

Vanucchi, G. 2007. *Problemi di geotecnica nel restauro monumentale*. Master di 2° Livello, Restauro, Protezione e Sicurezza degli Edifici Storici e Monumentali. Academic Year 2006–07, Universitá di Firenze.

Wood, C. and Oreszczyn, T. 2004. *Building Regulations and Historic Buildings. Balancing the needs for energy conservation with those of building conservation: an Interim Guidance Note on the application of Part L*. Swindon: English Heritage.

Woods, E. 2004. Bored tunnels. *The Arup Journal* 1: 22–28.

Yeomans, D. T. 1985. History of timber structures. In J. Sunley and B. Bedding (eds) *Timber in Construction*. London: TRADA.

Yeomans, D. T. 1999. The development of timber as a structural material. *Series: Studies in the History of Civil Engineering*; vol. 8. Aldershot: Ashgate/Variorum.

Yeomans, D. T. 2003. *The Repair of Historic Timber Structures*. London: Thomas Telford.

Zimmermann, T., Strauss, A. and Bergmeister, K. 2010. Numerical investigation of historic masonry walls under normal and shear load. *Construction and Building Materials*, 24: 1385–1391.

Case studies and applications

AAVV 2000. *Museo Nazionale Romano: Crypta Balbi*. Ministero per i Beni e le Attività Culturali, Soprintendenza Archeologica di Roma. Milan: Electa.

Alonso del Val, M. A. and Hernandez Minguillón, R. J. 2002. *Rehabilitacion de la antigua biblioteca del Monasterio de Fitero*. Available at: www.ahasociados.com (accessed September 2011).

Armaly, M., Blasi, C. and Hannah, L. 2004. Stari Most: rebuilding more than a historic bridge in Mostar. *Museum International*, 56(4): 6–17.

Arquimatica 2011a. *Nueva sede del Ayuntamiento de Madrid en el edificio de Correos*. Available at: www.arquimatica.com/xhtml/proyecto.php?id_prj=11&id_news=1 (accessed September 2011).

Arquimatica 2011b. *Discussion about the design and construction of the roof in Palacio de Cibeles*. Interview by Dimitris Theodossopoulos, Madrid, March 2011.

Arsène, A. 1918. *Les monuments français détruits par l'Allemagne*. Paris and Nancy: Berger-Levrault, Libraires-Éditeurs.

Avanti Architects 2009. *St Peter's Seminary, Cardross: Conservation Assessment*. A report prepared for HS. Available at: www.historic-scotland.gov.uk/index/news/indepth/stpeters/stpeters-avantireport.htm (accessed August 2011).

Azkarate, A., Cámara, L., Lasagabater, J. I. and Latorre, P. 2001. *Plan Director para la restauración de la catedral de Santa María de Vitoria-Gasteiz*. Vitoria: Foundation Catedral Santa Maria.

Barber, J. 2009. Discussion of broch construction. *Towards An Improved Understanding of the Engineering of Brochs, 2009*, See http://sites.google.com/site/brochgroup/home for further details (accessed March 2011).

Barthel, R., Maus, H. and Jagfeld, M. 2006. Artistry and ingenuity of Gothic vaults at the example of St. Georg in Nördlingen. In: *2nd International Congress in Construction History*, Cambridge University, 29 March–2 April. London: Construction History Society.

Beard, A. 2001. A future for Park Hill. In S. Macdonald (ed.) *The Care and Conservation of Mid-Twentieth Century Architecture*. Donhead St Mary: Donhead Publishing.

Bilson, J. 1906. Amiens Cathedral and Mr. Goodyear's 'refinements': a criticism. *Journal of the Royal Institute of British Architects*, 3rd series, XIII(15): 397–417.

Bizley, G. 2009. Bringing a smooth finish to a turbulent history. *Concrete Quarterly*, no. 227: 4–7.

Bosco, N. 2000. Il controllo della luce. Restauro di Palazzo Carignano, Torino (1982–85). In N. Bosco, *Andrea Bruno, tecniche esecutive e dettagli progettuali*. Milan: Libreria CLUP.

Brown, S. 2005. Millennium and beyond. *Structural Engineer*, 83(20): 34–42.

Brownlie, K. 1999. Naval glazing. *Architects' Journal*, 6 May 1999.

Burland, J. 2003. A tale of two towers: Big Ben and Pisa. *ICE Annual Conference*.

Cantacuzino, S. 1975. *New Uses for Old Buildings*. London: Architectural Press.

Carbonara, G. 2005. *Atlante del restauro*. Turin: UTET.

Castigliano, A. 1888. *Manuale pratico per gli ingegneri*. Turin: A. F. Negro.

Cecchini, L. 2004. Il Ponte Pietra e il Ponte di Castelvecchio prima e dopo la Seconda Guerra Mondiale. *Conference 'De Verona Eiusque Aquis'*, Centrum Latinitatis Europae (CLE), Verona, 3 May.

Ceradini, V. 1993. Analisi stereotomica per lo studio della stabilità della facciata di San Carlino alle Quattro Fontane, in *San Carlino alle Quattro Fontane. Il restauro della facciata. Note di cantiere*. Rome: N.M. Gammino.

Ceradini, V. 1996. Modelli sperimentali di volte. *National Conference La meccanica delle murature tra teoria e progetto*, Messina, pp. 157–166.

Clifton, G. and Willis, G. 2007. Dean's Eye Window – the reconstruction of a medieval rose window at Lincoln Cathedral. *The Structural Engineer*, 85(3): 40–46.

Connell, G. S. 1993. Restoration of Brunel's Paddington Station roof. *Proceedings of the ICE – Civil Engineering*, 97(1): 10–18.

Cooper, M. 2010. Northern verve. Structures special: Wigan Life Centre. *New Civil Engineer*, 2 September: 22–23.

Coste, A. 1995. Le calcul par la méthode des éléments finis appliqué à la restauration. Une expérience: la cathédrale de Beauvais. In E. Benvenuto and P. Radelet-de-Grave (eds) *Entre mécanique et architecture*. Basel: Birkhäuser Verlag.

Cramer, J. and Breitling, S. 2007. *Architecture in Existing Fabric*. Basel: Birkhäuser.

Croci, G. 1998. Gli interventi definitivi per il consolidamento. In G. Basile and P. N. Giandomenico (eds) *La Basilica di San Francesco in Assisi. Progetto di restauro, primi avanzamenti. Quaderno 3, June 1998*. Sacro Convento di San Francesco in Assisi.

Croci, G. 2000. General methodology for the structural restoration of historic buildings: the cases of the Tower of Pisa and the Basilica of Assisi. *Journal of Cultural Heritage*, 1: 7–18.

Croci, G. and Scoppola, F. 1993. Il restauro di Palazzo Altemps. *TeMA Tempo Materia e Architettura, Rivista quadrimestrale di restauro*, 1: 11–25.

Davey, P. 2007. Diocesan dialogue. Zumthor in Cologne. *Architectural Review*, no. 1329: 36–41.

De Matteis, G. F. and Mazzolani, F. M. 2010. The Fossanova Church: seismic vulnerability assessment by numeric and physical testing. *International Journal of Architectural Heritage*, 4(3): 222–245.

Del Monti, C. 2004. L'aula dell'Oratorio dei XL Martiri. Criteri e metodologia del restauro. In J. Osborne, J. Rasmus Brandt and G. Morganti (eds) *Santa Maria Antiqua al Foro Romano cento anni dopo*. Rome: Campisano Editore, pp. 153–165.

Dirckinck-Holmfeld, K. 1998. Il Castello di Kolding in Danimarca: conservazione e innovazione nell'intervento di Inger e Johannes Exner (1972–1991). ANAΓKH (ÁNANKE), no. 22, June: 42–67 (translated from *Architektur DK*, 1, 1994).

Dorset County Council. 2007. *Dorchester Roman Town House Conservation Plan*.

Fawcett, A. P., Barks, J., Walker, D. and Webb, D. 1994. Factory with a facelift. Boots factory in Nottingham. *Architects' Journal*, 3 November: 31–41.

Fawcett, R. (ed.) 2001. *Stirling Castle: The Restoration of the Great Hall*. York: Council for British Archaeology.

Feldman, G. C. 2005. Fallingwater is no longer falling. *STRUCTURE Magazine*, September: 46–50.

Ferrari, M. 2009. Una teca che restituisce come una sineddoche un mondo finito. Museo e centro visitatori della Villa Romana di La Olmeda, Palencia, Spagna. *Casabella*, November: 42–51.

Fintikakis, N. (Φιντικακης) 2004. The new bioclimatic roof at Akrotiri, Santorini. *Architecktones*, no. 46: 30–33. (Originally published in Greek: Το νέο βιοκλιματικό στέγαστρο στο Ακρωτήρι Σαντορίνης. Αρχιτεκτονες.

formTL 2011. Hagar Qim and Mnajdra Temples Malta: membrane roofs. Available at: www.form-tl.de/en/projects/structures-buildings/hagar-qimmnajdra-malta.html (accessed November 2011).

Gallegos, G. and González, P. 2007. Rehabilitación de la iglesia de San Agustín como archivo municipal. Valladolid. *ON Diseño*, no. 281: 190–203.

García García, M. 2008. Intervenciones en la Villa Astur Romana de Veranes en Gijón. *Liño, Revista Annual de Historia del Arte*, 14: 211–216.

Gil Cornet, L. 2004. El Palacio Real de Olite: crònica de una obstinación. *Papeles del Partal*, no. 2, November: 121–153.

Gilman, R. 1920. The theory of Gothic architecture and the effect of shellfire at Rheims and Soissons. *American Journal of Archaeology*, 2nd series, 24(1): 37–72.

Giuffré, A. 1988. Restauro e sicurezza in zona sismica; la Cattedrale di Sant'Angelo dei Lombardi. *Palladio, Rivista di Storia dell'Architettura e Restauro*, 1st. Poligrafico e Zecca dello Stato, Rome, Italy, n. s., anno I[2].

Gold, R. and Knox, U. 2005. The rotating scaffolding in the dome of St Paul's Cathedral. *The Structural Engineer*, 6 December: 46–52.

González-Longo, C. 2008. Conserving, reinstating and converting Queensberry House. In B. Sinha and L. Tanacan (eds) *8th International Seminar on Structural Masonry*, Istanbul, 5–7 November, pp. 431–440.

González-Longo, C. and Theodossopoulos, D. 2009. The platform of the Temple of Venus and Rome. *Third International Congress on Construction History*, Cottbus, 20–24 May. Chair of Construction History and Structural Preservation of the Brandenburg University of Technology.

Hansford, M. 2002. Savings in the frame. 20–32 Baker Street. *New Civil Engineer*, 14 March: 16–17.

Hansford, M. 2008. Under-designed gusset plates to blame for Minneapolis bridge collapse: official investigation concludes. *New Civil Engineer*, 15 November.

Hirst, J. G., Wainwright, F. and Shaw, J. M. 2000. Tate Modern. *The Structural Engineer*, 78(4): 15–21.

Hoh-Slodczyck, C. 1999. The rehabilitation of the Einstein Tower: a careful and economical approach to historical building fabric. *DETAIL*, 7: 1257–1260.

Holton, A. 2006. The working space of the medieval master mason: the tracing houses of York Minster. *2nd International Congress in Construction History*, Cambridge University, 29 March–2 April 2006. London: Construction History Society.

Ibañez, F. 2010. Baños Árabes, Baza (Granada). *Arquitectura Viva Monografías*, nos. 141–142: 158–163.

Ioannidou, M. 2006. Seismic action on the Acropolis monuments. *The Acropolis Restoration News*, no. 6, July.

Jolly, C. K. and Brown, K. J. 1999. Prospects for the Prospect Room: the proposed strengthening of an Elizabethan timber floor at Wollaton Hall, Nottingham. *The Structural Engineer*, 77(13): 25–34.

Karanasos, K. 2006. Restoring again the Ionic columns of the Propylaia. *The Acropolis Restoration News*, no. 6, July.

Kidson, P. 1986. St Hugh's Choir, in *Medieval Art and Architecture at Lincoln*

Cathedral. Conference Transactions of the British Archaeological Association. Oxford: Oxbow Books, pp. 29–42.

Kouris, S. and Karaveziroglou-Weber, M. 2009. Research on the seismic strengthening of a medieval masonry campanile. *The Structural Engineer*, 87(7): 20–24.

Kulturhaus Schloss. 2010. *Bundestransferstelle Städtebaulicher Denkmalschutz: Gute Beispiele*. Available at: www.bbsr.bund.de/BBSR/DE/Stadtentwicklung/ Staedtebaufoerderung/StaedtebaulicherDenkmalschutz/Begleitforschung/ GuteBeispiele/GuteBeispiele.html (accessed February 2011).

Lawrence, S., Price, P. and Power, K. 1998. A new roof at Scotland's Stirling Castle. *Timber Framing 47*, 4 March: 4–8.

Lewis, D., Johnstone, P., Hadden, D., Wilkinson, R. and Tweedie, A. E. 2006. The Scottish Parliament, Holyrood building. *The Structural Engineer*, 84(5): 29–35.

Lorenz, S. 1964. Utilisation des edifices monumentaux en Pologne. *Il monumento per l'uomo*. Second International Congress of Restoration, Venice, 25–31 May.

Lorenzetti, M. 2000. *Museo Nazionale Romano Crypta Balbi, Roma*. Available at: www. architettiroma.it/progetti/p00494.aspx (accessed 1 November 2010).

Macdonald, S. 2001. *Preserving Post-War Heritage: The Care and Conservation of Mid-Twentieth Century Architecture*. Donhead St Mary: Donhead Publishing.

Manacorda, G. (ed.) 2004. *Crypta Balbi. Archeologia e storia di un paessaggio urbano*. Milan: Electa.

Mangado Beloqui, F. 2010. El Museo de Bellas Artes de Oviedo. *Liño, Revista annual de Historia del Arte*, no. 16: 179–186.

Manos, G. C., Soulis, V. J. and Diagouma, A. 2008. Numerical investigation of the behaviour of the church of Agia Triada, Drakotrypa, Greece. *Advances in Engineering Software*, 39: 284–300.

Martín, G. 2010. A fondo: Remodelación del Palacio de Cibeles. Estudio Arquimática. *Blog Proyectosinergias*, 10 October. Available at: www.proyectosinergias. com/2010/10/fondo-remodelacion-del-palacio-de.html (accessed September 2011).

Martínez, J. L., Martin-Caro, J. A., Torrico, J. and León, J. L. 2000. The 'Silla de la Reina' tower in the Cathedral of León. In J. Adell (ed.) *Proceedings of the 12th International Brick/Block Masonry Conference*, Madrid.

Maunder, E. W. A. 1995. Some structural studies of Exeter Cathedral. *The Structural Engineer*, 73(4): 105–110.

McDowell, A., Sedgwick, A. and Lenczner, A. 1994. Shedding light on the Louvre. *The Architects' Journal* (Technical & Practice), 27 April: 29–31.

Meade, E. P. and Hudson, N. R. 2010. Structural concrete repairs to the Solomon R. Guggenheim Museum, New York, New York, US. In M. C. Forde (ed.) *International Conference: Structural Faults and Repair*, Edinburgh, June 2010.

Menéndez Pidal, L. 1960. La Cámara Santa, su destrucción y reconstrucción. *Boletin del Instituto de Estudios Asturianos*, 39(1): 3–34. Oviedo: Instituto de Estudios Asturianos.

Meynell, M. 2009. *An Historical Guide to All Souls Church*. London: All Souls.

Miltiadou-Fezans, A. 2008. A multi-disciplinary approach for the structural restoration of the Katholikon of Dafni Mpnastery in Attica Greece. In D. F. D'Ayala and E. Fodde (eds) *Structural Analysis of Historic Construction*. London: Taylor & Francis.

Minissi, F. 1961. Protection of the mosaic pavements of the Roman Villa at Piazza Armerina (Sicily). *Museum*, XIV(2): 128–132.

Murphy, K. 2002. The Villa Savoie and the Modernist historic monument. *Journal of the Society of Architectural Historians*, 61(1) March: 68–89.

Murphy, R. 2011. National Museum of Scotland, Edinburgh, by Gareth Hoskins Architects. *Architects' Journal*, 1 September.

Orlandi, D. 2011. *L'accessibilità dei siti archeologici: il Colosseo*. Superabile, a 'Contact Center Integrato' by INAIL. Available at: www.superabile.it/web/it/CANALI_TEMATICI/Senza_Barriere/Soluzioni_Progettuali/Spazi_ed_edifici/info-1659734719.html (accessed September 2011).

Owen, E. 2007. Piling on the style. Structures: 80 Mosley Street. *New Civil Engineer*, 3 May: 22–23.

Papadopoulos, K. 2010. The restoration of the north-side foundation of the Temple of Apollo Epikourios. *International Journal of Architectural Heritage*, 4(3): 246–269.

Pardo Calvo, F. and García Tapia, B. 2004. Reforma y ampliación del Museo Arqueológico de Asturias. *Revista museos.es nº 0*, pp. 118–131. Dirección General de Bellas Artes y Bienes Culturales del Ministerio de Cultura.

Poleni, G. 1748. *Memorie istoriche della Gran Cupola del Tempio Vaticano*. Padova: Edizioni del Seminario.

Porras, I. 2011. Madrid inaugura el renovado Palacio de Cibeles y lo abre a los ciudadanos. *Diario Design*, 27 April. Available at: http://diariodesign.com/2011/04/madrid-inaugura-el-renovado-palacio-de-cibeles-y-recupera-el-proyecto-original/ (accessed September 2011).

Pörtner, R. 1992. Verformungen des Traggefüges und Schäden im Dormentbau Kloster Maulbronn. *Jahrbuch 1990; SFB 315*. Berlin: Ernst & Sohn.

Powell, K. and Schollar, T. 1994. Restoring a milestone of Modernism. Building study: De La Warr Pavilion. *Architects' Journal*, 16 February: 35–44.

Psycharis, I., Lemos, J. V., Papastamatiou, D., Zambas, C. and Papantonopoulos, C. 2003. Numerical study of the seismic behaviour of a part of the Parthenon Pronaos. *Earthquake Engineering and Structural Dynamics*, 32: 2063–2084.

Romagnoli, L. 2010. *Parco Archeologico di Fregellae*. In Archiportale, 26 February. Available at: www.archiportale.com/progetti/laura-romagnoli/arce/parco-archeo-logico-di-fregellae_27243.html (accessed October 2010).

Rowson, J. 2008. Bearing the load. Temporary works: Regent Street. *New Civil Engineer*, 24 April: 26–28.

Rowson, J. 2009. Southend Pier: born again. *New Civil Engineer*, 28 May 2009.

Salvo, S. 2010. Il recupero delle facciate continue in metallo e vetro e il 'caso' del grattacielo Pirelli di Milano. *International Seminar: Glass in 20th Century Architecture: Preservation and Restoration*. Accademia di architettura: Università della Svizzera Italiana, 16–7 November.

Sanchez de Rio, I. 1928. El cuarto Depósito de aguas, de Oviedo. *Revista de Obras Públicas*, Year LXXVI, no. 2, 1 August: 506.

Schittich, C. 2003. *Building in Existing Fabric: Refurbishment, Extensions, New Design*. Munich: Edition Detail; Basel: Birkhäuser.

Silman, R. and Matteo, J. 2001. Repair and retrofit. Is Fallingwater falling down? *Structure*, July/August.

Smith, M. 1999. The Engineering building, University of Leicester fire safety works. *Context, magazine of the Institute of Historic Building Conservation*, no. 63, September.

Stratton, M. 1997. *Structure and Style: Conserving 20th Century Buildings*. London: Taylor & Francis.

Sutherland, A. and Thew, I. 2010. Structural behaviour and technology of Iron-age brochs. Dissertation for the degree of MEng Structural Engineering with Architecture, University of Edinburgh.

Tanoulas, T. 2006. The modern marble copies of the Ionic column capitals of the Propylaia. *The Acropolis Restoration News*, no. 6, July.

Taupin, J. L. 1993. Cathédrale de Beauvais: de l'incertitude à la decision. In G. Croci (ed.) *Symposium 'Structural Preservation of the Architectural Heritage'*, Rome, 15–17 September. Zurich: IABSE.

Tensinet 1990. Roof over Archaeological Excavation Sites on Pianosa and in Desenzano sul Garda (Italy). Available at: www.tensinet.com/database/viewProject/3820 (accessed November 2008).

Theodossopoulos, D. 2004. Structural scheme of the cathedral of Burgos. In C. Modena, P. B. Lourenço and P. Roca (eds) *International Conference. Structural Analysis of Historical Constructions* (SAHC), Padua 10–13 November. London: Taylor & Francis, pp. 643–652.

Theodossopoulos, D. 2008. Structural design of High Gothic vaulting systems in England. *International Journal of Architectural Heritage*, 2(1): 1–24.

Theodossopoulos, D., Sinha, B. P. and Usmani, A. S. 2003. Case study of the failure of a cross vault: Church of Holyrood Abbey. *Journal of Architectural Engineering ASCE*, 9(3): 109–117.

Truman, N. 1946. *Holme by Newark Church and its Founder*. Gloucester and London: British Publishing Company.

Vivio, B. 2010. *Franco Minissi: musei e restauri. La trasparenza come valore*. Rome: Gangemi.

Von Gerkan, M. 1991. *Von Gerkan, Marg und Partner: architecture, 1988–1991*. Basel, and Boston: Birkhäuser.

Warszawa 1939 (2011). Architektura Przedwojennej Warszawy. Fundacja 'Warszawa1939.pl'. Available at: www.warszawa1939.pl/ (accessed July 2011).

Wenzel, F. (ed.) 2007. *Berichte vom Wiederaufbau der Frauenkirche zu Dresden*. Karlsruhe: Universitatsverlag Karlsruhe.

Yárnoz, J. and Yárnoz, J. 1925. La restauración del Palacio Real de Olite. *Boletín de la Comisión de Monumentos Histórico-Artísticos de Navarra*, no. 63, Tomo XVI, Tercer Trimestre. Pamplona, pp. 315–376.

Yeomans, D. and Cottam, D. 1989. An architect/engineer collaboration: the Tecton Arup flats. *The Structural Engineer*, 67(10): 183–188.

Index